Building Construction

AUTHORS

Dr. Rinku Kumar
Assistant Professor, Lecturer,
Government Polytechnic Mankeda,
Agra, Uttar Pradesh

Dr. Sandeep Panchal
Lecturer,
Government Polytechnic Mankeda,
Agra, Uttar Pradesh

REVIEWER

Mrs. S. Santhi
Lecturer (Sr. Grade),
A.M.K. Technological Polytechnic College,
Thiruvallur, Tamil Nadu

KHANNA BOOK PUBLISHING CO. (P) Ltd.
4C/4344, Ansari Road, Darya Ganj,
New Delhi, India

All India Council for Technical Education
Nelson Mandela Marg, Vasant Kunj,
New Delhi, India

Dear Readers,

To prevent the piracy, this book is secured with HIGH SECURITY HOLOGRAM on the front title cover. In case you don't find the hologram on the front cover title, please write us to at contact@khannabooks.com or whatsapp us at +91-99109 09320 and avail special gift voucher for yourself.

Specimen of Hologram on front Cover title:

Moreover, there is a SPECIAL DISCOUNT COUPON for you with EVERY HOLOGRAM.

How to avail this SPECIAL DISCOUNT:

Step 1: Scratch the hologram

Step 2: Under the scratch area, your "coupon code" is available

Step 3: Logon to www.khannabooks.com

Step 4: Use your "coupon code" in the shopping cart and get your copy at a special discount

Step 5: Enjoy your reading!

ISBN: 978-93-5538-739-4
Book Code: DIP436EN

Building Construction

by Dr. Rinku Kumar and Dr. Sandeep Panchal

First Edition: 2025

Originally Published by:
All India Council for Technical Education, Ministry of Education, GOI

Exclusively Published & Distributed by:
Khanna Book Publishing Co. (P) Ltd.
4C/4344, Ansari Road, Daryaganj,
New Delhi (INDIA)

Visit us at: **www.khannabooks.com**
Write us at: **contact@khannabooks.com**

To view complete list of books, Please scan the QR Code:

KPH

Printed in Bharat.

प्रो. म. जगदीश कुमार
अध्यक्ष
Prof. M. Jagadesh Kumar
Chairman

सत्यमेव जयते

आज़ादी का अमृत महोत्सव

अखिल भारतीय तकनीकी शिक्षा परिषद्
(भारत सरकार का एक सांविधिक निकाय)
(शिक्षा मंत्रालय, भारत सरकार)
नेल्सन मंडेला मार्ग, वसंत कुंज, नई दिल्ली–110070
दूरभाष : 011-26131498
ई–मेल : chairman@aicte-india.org

ALL INDIA COUNCIL FOR TECHNICAL EDUCATION
(A STATUTORY BODY OF THE GOVT. OF INDIA)
(Ministry of Education, Govt. of India)
Nelson Mandela Marg, Vasant Kunj, New Delhi-110070
Phone : 011-26131498
E-mail : chairman@aicte-india.org

FOREWORD

Engineers are the backbone of the modern society. It is through them that engineering marvels have happened and improved quality of life across the world. They have driven humanity towards greater heights in a more evolved and unprecedented manner.

The All India Council for Technical Education (AICTE), led from the front and assisted students, faculty & institutions in every possible manner towards the strengthening of the technical education in the country. AICTE is always working towards promoting quality Technical Education to make India a modern developed nation with the integration of modern knowledge & traditional knowledge for the welfare of mankind.

An array of initiatives have been taken by AICTE in last decade which have been accelerate now by the National Education Policy (NEP) 2020. The implementation of NEP under the visionary leadership of Hon'ble Prime Minister of India envisages the provision for education in regional languages to all, thereby ensuring that every graduate becomes competent enough and is in a position to contribute towards the national growth and development through innovation & entrepreneurship.

One of the spheres where AICTE had been relentlessly working since 2021-22 is providing high quality books prepared and translated by eminent educators in various Indian languages to its engineering students at Under Graduate & Diploma level. For the second year students, AICTE has identified 87 books at Under Graduate and Diploma Level courses, for translation in 12 Indian languages - Hindi, Tamil, Gujarati, Odia, Bengali, Kannada, Urdu, Punjabi, Telugu, Marathi, Assamese & Malayalam. In addition to the English medium, the 1044 books in different Indian Languages are going to support to engineering students to learn in their mother tongue. Currently, there are 39 institutions in 11 states offering courses in Indian languages in 7 disciplines like Biomedical Engineering, Civil Engineering, Computer Science & Engineering, Electrical Engineering, Electronics & Communication Engineering, Information Technology Engineering & Mechanical Engineering, Architecture, and Interior Designing. This will become possible due to active involvement and support of universities/institutions in different states.

On behalf of AICTE, I express sincere gratitude to all distinguished authors, reviewers and translators from different IITs, NITs and other institutions for their admirable contribution in a very short span of time.

AICTE is confident that these out comes based books with their rich content will help technical students master the subjects with factor comprehension and greater ease.

(Prof. M. Jagadesh Kumar)

Prof. M. Jagadesh Kumar
Chairman

FOREWORD

(Prof. M. Jagadesh Kumar)

ACKNOWLEDGEMENT

The authors are grateful to the authorities of AICTE, particularly Prof. M. Jagadesh Kumar, Chairman; Prof. M. P. Poonia, Vice-Chairman; Prof. Rajive Kumar, Member-Secretary and Dr Amit Kumar Srivastava, Director, Faculty Development Cell for their planning to publish the books on Building Construction. We sincerely acknowledge the valuable contributions of the reviewer of the book Mrs. S. Santhi, Lecturer (Sr. Grade), Civil Engineering, A. M. K. Technological Polytechnic College, Thiruvallur, Tamil Nadu for making it students' friendly and giving a better shape in an artistic manner.

We would like to offer special thanks to Mr. Anil Kumar, Mrs. Poonam Kumari, Mr. Vishwnath Pratap Singh, Dr. Sudhanshu Singh, Dr. Apoorva Saxena, Shri Krishna, Mr. Aman Singh Bhadouria, Mr. Chander Pal, Mr. Alaxender Panchal, Mrs. Kiran, Mrs. Kareena, Ms. Shivani, and Mr. Dharmendra Singh to provide moral support, encouragement and helped us in every possible way.

Last, but definitely not the least, we would like to express our deepest thanks to ours parents and wives (Dr. Rinku Kumar-Mrs. Sandhya Singh; Dr. Sandeep Panchal-Mrs. Anjali) who helped us in every possible way. They have always taken all the pains to provide us with the best throughout our lives. We have reached this position today only because of them. Their unconditional love, care, support and constant encouragement allowed us to finish our book successfully.

This book is an outcome of various suggestions of AICTE members, experts, and authors who shared their opinion and thought to further develop engineering education in our country. Acknowledgements are due to the contributors and different workers in this field whose published books, review articles, papers, photographs, footnotes, references and other valuable information enriched us at the time of writing the book.

Dr. Rinku Kumar

Dr. Sandeep Panchal

PREFACE

The book titled "Building Construction" is an outcome of the experience of our teaching of building construction and building materials and research work carried out at IIT Roorkee, NIT Hamirpur, and Delhi Technological University. The book is written to provide the knowledge of basics of building construction to civil engineering students and professionals. Keeping in mind the purpose of wide coverage as well as to provide essential supplementary information, we have included the topics recommended by AICTE, in a very systematic and orderly manner throughout the book. Efforts have been made to explain the fundamental concepts of the subject in the simplest possible way.

During the process of preparation of the manuscript, we considered the various standard textbooks. As the subject is more about visualization, we have included the 2-D and 3-D diagrams for a better explanation of the concepts. The book covers all types of details of building construction requires for diploma-level students and these have been presented in a very logical and systematic manner.

Apart from illustrations and examples as required, we have enriched the book with numerous unsolved questions in every unit for a proper understanding of the related topics. We have tried to include questions that will make the students think about the problem. It will help in boosting the problem-solving skills of the students. The book is written in a way that the students will find interesting to read. The information provided in the chapters will build a strong curiosity about the new building construction techniques in the students. In addition, some interesting historical information for the users is provided under the heading "Know more". We have provided some details of Indian standard codes and 3-D view of building components in the appendix and annexure section.

As far as the present book is concerned, "Building Construction" is meant to provide a thorough grounding in building construction on the topics covered. This part of the building construction book will prepare engineering students to apply their knowledge of building construction to tackle the problems that they will face in the civil engineering field. The subject matters are presented in a constructive and simple manner so that an engineering diploma prepares students to work in the field in an efficient and creative manner. This book will help the civil engineers and professionals in their fieldwork too.

We sincerely hope that the book will inspire the students to learn and discuss the ideas behind basic principles of building construction and will surely contribute to the development of a solid foundation of the subject. We would be thankful to all beneficial comments and suggestions which will contribute to the improvement of future editions of the book. It gives us immense pleasure to place this book in the hands of the teachers, students, and civil engineering professionals. It was indeed a big pleasure to work on different aspects covering in the book.

Dr. Rinku Kumar

Dr. Sandeep Panchal

OUTCOME BASED EDUCATION

For the implementation of an outcome-based education the first requirement is to develop an outcome based curriculum and incorporate an outcome based assessment in the education system. By going through outcome based assessments evaluators will be able to evaluate whether the students have achieved the outlined standard, specific and measurable outcomes. With the proper incorporation of outcome based education there will be a definite commitment to achieve a minimum standard for all learners without giving up at any level. At the end of the programme running with the aid of outcome based education, a student will be able to arrive at the following outcomes:

PO-1. Engineering knowledge: Apply the knowledge of mathematics, science, engineering fundamentals, and an engineering specialization to the solution of complex engineering problems.

PO-2. Problem analysis: Identify, formulate, review research literature, and analyze complex engineering problems reaching substantiated conclusions using first principles of mathematics, natural sciences, and engineering sciences.

PO-3. Design / development of solutions: Design solutions for complex engineering problems and design system components or processes that meet the specified needs with appropriate consideration for the public health and safety, and the cultural, societal, and environmental considerations.

PO-4. Conduct investigations of complex problems: Use research-based knowledge and research methods including design of experiments, analysis and interpretation of data, and synthesis of the information to provide valid conclusions.

PO-5 Modern tool usage: Create, select, and apply appropriate techniques, resources, and modern engineering and IT tools including prediction and modeling to complex engineering activities with an understanding of the limitations.

PO-6. The engineer and society: Apply reasoning informed by the contextual knowledge to assess societal, health, safety, legal and cultural issues and the consequent responsibilities relevant to the professional engineering practice.

PO-7. Environment and sustainability: Understand the impact of the professional engineering solutions in societal and environmental contexts, and demonstrate the knowledge of, and need for sustainable development.

COURSE OUTCOMES

After By the end of the course the students are expected to learn:

CO-1: To identify various components of building structures

CO-2: To propose suitable type of foundation for building structures

CO-3: To select suitable type of masonry for building structures

CO-4: To propose relevant means of communications for different types of buildings

CO-5: To select relevant material for finishing works

Course Outcomes	EXPECTED MAPPING WITH PROGRAMME OUTCOMES (1- Weak Correlation; 2- Medium correlation; 3- Strong Correlation)						
	PO-1	PO-2	PO-3	PO-4	PO-5	PO-6	PO-7
CO-1	2	1	1	1	2	-	-
CO-2	3	3	3	1	2	1	-
CO-3	3	2	3	1	1	1	-
CO-4	3	2	2	1	1	-	-
CO-5	3	1	1	1	-	-	-

GUIDELINES FOR TEACHERS

To implement Outcome Based Education (OBE) knowledge level and skill set of the students should be enhanced. Teachers should take a major responsibility for the proper implementation of OBE. Some of the responsibilities (not limited to) for the teachers in OBE system may be as follows:

- Within reasonable constraint, they should manoeuvre time to the best advantage of all students.
- They should assess the students only upon certain defined criterion without considering any other potential ineligibility to discriminate them.
- They should try to grow the learning abilities of the students to a certain level before they leave the institute.
- They should try to ensure that all the students are equipped with the quality knowledge as well as competence after they finish their education.
- They should always encourage the students to develop their ultimate performance capabilities.
- They should facilitate and encourage group work and team work to consolidate newer approach.
- They should follow Blooms *taxonomy in every part of the assessment.*

Bloom's Taxonomy

Level	Teacher should Check	Student should be able to	Possible Mode of Assessment
Creating	Students ability to create	Design or Create	Mini project
Evaluating	Students ability to Justify	Argue or Defend	Assignment
Analysing	Students ability to distinguish	Differentiate or Distinguish	Project/Lab Methodology
Applying	Students ability to use information	Operate or Demonstrate	Technical Presentation/ Demonstration
Understanding	Students ability to explain the ideas	Explain or Classify	Presentation/Seminar
Remembering	Students ability to recall (or remember)	Define or Recall	Quiz

GUIDELINES FOR STUDENTS

Students should take equal responsibility for implementing the OBE. Some of the responsibilities (not limited to) for the students in OBE system are as follows:

- Students should be well aware of each UO before the start of a unit in each and every course.
- Students should be well aware of each CO before the start of the course.
- Students should be well aware of each PO before the start of the programme.
- Students should think critically and reasonably with proper reflection and action.
- Learning of the students should be connected and integrated with practical and real life consequences.
- Students should be well aware of their competency at every level of OBE.

ABBREVIATIONS AND SYMBOLS

List of Abbreviations

General Terms			
Abbreviations	**Full form**	**Abbreviations**	**Full form**
SP	Special Publication	OPC	Ordinary Portland cement
3D	Three Dimensional	GL	Ground level
LPG	Liquefied Petroleum Gas	RCC	Reinforced cement concrete
2D	Two Dimensional	ER	Excavation line wooden peg
CNG	Compressed Natural Gas	CL	Centre line of foundation marking peg
PVC	Polyvinyl chloride	CPWD	Central public work department
IS	Indian Standard	GI	Galvanized Iron
R	Reference wooden peg	RCC	Ultraviolet Excavation line
DPC	Damp Proof Course	POP	Plaster of Paris
AC	Asbestos Cement		
FRP	Fiber-reinforced Polymer		

List of Units and Symbols

Symbols/ Units	Description	Symbols/ Units	Description
A-1 to A-9	Sub type of residential building	%	Percentage
B-1 to B-2	Sub type of educational building	gm	Gram
C-1 to C-3	Sub type of institutional building	cc	Cubic centimeters
D-1 to D-7	Sub type of assembly building	P	Load
E-1 to E-5	Sub type of business building	γ	Unit weight of the soil
F-1 to F-3	Sub type of mercantile building	φ	Angle of repose
G-1 to G-3	Sub Type of industrial building	kN	Kilo newton
cm	Centimetre	m^3	Cubic metre
MPa	Mega Pascal	q_s	Safe bearing capacity of the soil
m^2	Square metre	n	Spread in horizontal with respect to vertical
m	metre	q	Net bearing capacity of soil
°	degree	a	Offset of concrete

Symbols/ Units	Description	Symbols/ Units	Description
NBC	National building code	B_3	One quarter bat
mm	millimeter	L	Leveler
Kg	Kilogram	S	Stretcher
H	Header	Q	Queen closer
B_1	Three-quarter bat	B_2	Half bat
N	Newton		

LIST OF FIGURES

1. Overview of Building Components

Fig. 1.1: Load-bearing structure 6
Fig. 1.2: Framed structure 6
Fig. 1.3: Substructure and its components 8
Fig. 1.4: Components of the superstructure 9
Fig. 1.5: Cavity wall 10
Fig. 1.6: Floors of a house 11
Fig. 1.7: Mezzanine Floor 12
Fig. 1.8: Structural Members of the Building 13

2. Construction of Substructure

Fig. 2.1: Job layout of building/mall under construction 19
Fig. 2.2: Job layout by center line method 21
Fig. 2.3: Job layout by face line method 22
Fig. 2.4 (a): Stray bracings 25
Fig. 2.4 (b): Box sheeting 25
Fig 2.4. (c): Runner system 25
Fig. 2.4 (d): Vertical sheeting 26
Fig. 2.5: Hand tools for excavation (a) Rake, (b) Basket, (c) Hoe or phaorah, (d) Rammer, (e) Shovel 28
Fig. 2.6 (a): Tracked excavator 29
Fig. 2.6 (b): Backhoe 29
Fig. 2.7 (a): Single column footing 31
Fig. 2.7 (b): Plan of single footing 31
Fig. 2.7 (c): Single column footing 32
Fig. 2.7 (d): Details of plan of single footing 32
Fig. 2.8 (a): Stepped column footing 32
Fig. 2.8 (b): Plan of stepped column footing 32
Fig. 2.8 (c): Elevation of stepped column footing 33
Fig. 2.9 (a): Sloped column footing 34
Fig. 2.9 (b): Plan of sloped column footing 34
Fig. 2.10 (a): Strip wall footing 34
Fig. 2.10 (b): Plan of strip wall footing 34
Fig. 2.11 (a): Stepped wall footing 35
Fig. 2.11 (b): Plan of stepped wall footing 35
Fig. 2.11 (c): Elevation of steps in stepped footing 36
Fig. 2.12 (a): Grillage foundation 37
Fig. 2.12 (b): Plan of grillage foundation 37
Fig. 2.13 (a): Combined rectangular footing 38

Fig. 2.13 (b): Plan of combined rectangular footing 38
Fig. 2.14 (a): Combined trapezoidal footing 38
Fig. 2.14 (b): Plan of combined trapezoidal footing 38
Fig. 2.15 (a): Rectangular column wall footing 39
Fig. 2.15 (b): Trapezoidal column wall footing 39
Fig. 2.16 (a): Strap footing 40
Fig. 2.16 (b): Plan of strap footing 40
Fig. 2.17 (a): Raft footing 42
Fig. 2.17 (b): Plan of raft footing 41
Fig. 2.18 (a): Pile foundation 42
Fig. 2.18 (b): Mechanism of end bearing piles 42
Fig. 2.19: Friction piles 42
Fig. 2.20: Combined action piles 43
Fig. 2.21 (a): Straight shaft pier 44
Fig. 2.21 (b): Bell-shaped pier 44
Fig. 2.22: Open caissons 45
Fig. 2.23: Floating caissons 45
Fig. 2.24: Pneumatic caisson 46
Fig. 2.25: Well point 47
Fig. 2.26: Single-stage well points 48
Fig. 2.27: Multi-stage well points 48
Fig. 2.28: Vacuum well points 49
Fig. 2.29: Deep well system 50
Fig. 2.30: Cofferdam 50
Fig. 2.31: Earth cofferdam 51
Fig. 2.32: Rockfill cofferdam 51
Fig. 2.33: Sheet pile cofferdam 52

3. Construction of Superstructure

Fig. 3.1: Terms used in stone masonry 58
Fig. 3.2: Dry rubble masonry 59
Fig. 3.3 (a): Uncoursed random rubble 60
Fig. 3.3 (b): Coursed random rubble 60
Fig. 3.4: Uncoursed square rubble masonry 61
Fig. 3.5 (a): Irregular coursed square rubble 61
Fig. 3.5 (b): Regularly coursed square rubble 61
Fig. 3.6: Details of flint wall 62
Fig. 3.7: Ashlar rough tooled masonry 63
Fig. 3.8: Ashlar fine tooled masonry 63
Fig. 3.9: Ashlar chamfered masonry 64
Fig. 3.10: Ashlar facing masonry 64

Fig. 3.11 (a): Joints in stone masonry 65
Fig. 3.11 (b): Typical joints in stone masonry 66
Fig. 3.12: Modular brick 68
Fig. 3.13 (a): King closer 68
Fig. 3.13 (b): Queen closer 68
Fig. 3.14 (a): Bevelled closer 68
Fig. 3.14 (b): Mitred closer 68
Fig. 3.15 (a): Half bat 69
Fig. 3.15 (b): Three-quarter bat 69
Fig. 3.15 (c): Bevelled bat 69
Fig. 3.16: Terms used in brick masonry 70
Fig. 3.17: Terms used in brick masonry 71
Fig. 3.18: Header bond 72
Fig. 3.19: English bond 73
Fig. 3.20 (a): Plan for one brick thick wall 73
Fig. 3.20 (b): Plan for one and half brick thick wall 74
Fig. 3.20 (c): Plan for two brick thick wall 74
Fig. 3.21: Elevation of Flemish bond 75
Fig. 3.22 (a): Plan for one brick thick wall 75
Fig. 3.22 (b): Plan for one and half brick thick wall 76
Fig. 3.22 (c): Plan for two brick thick wall 76
Fig. 3.23 (a): Plan for one and half brick thick wall 77
Fig. 3.23 (b): Plan for two brick thick wall 77
Fig. 3.24 (a): Tee-junction between half brick thick wall and one brick thick wall 78
Fig. 3.24 (b): Tee-junction between one and a half brick thick wall and one brick thick wall 79
Fig. 3.24 (c): Tee-junction between two one and a half brick thick walls 79
Fig. 3.24 (d): Tee-junction between one and a half brick thick wall and two brick thick wall 80
Fig. 3.25 (a): Tee-junction between the one brick thick internal wall in English bond and one brick thick external wall in Double Flemish bond 80
Fig. 3.25 (b): Tee-junction between the one and a half brick thick internal wall in English bond and Two brick thick external wall in Double Flemish bond 81
Fig. 3.26 (a): Tee junction between one brick thick wall and half brick thick wall in Flemish bond 81
Fig. 3.26 (b): Tee junction between one and a half brick thick wall and one brick thick wall in Flemish bond 82
Fig. 3.26 (c): Tee junction between one and a half brick thick wall and two brick thick wall in Flemish bond 82
Fig. 3.27 (a): Cross junction between one and a half brick thick wall and one brick thick wall 83
Fig. 3.27 (b): Cross junction between two one and a half brick thick walls 83
Fig. 3.27 (c): Cross junction between two one and a half brick thick wall and two brick thick wall 84
Fig. 3.28 (a): Squint junction between one brick thick wall and one and a half brick thick wall 84
Fig. 3.28 (b): Squint junction between one and half brick thick walls 85

Fig. 3.29 (a): Squint junction between one brick thick wall and one and a half brick thick wall 85
Fig. 3.29 (b): Squint junction between two one and a half brick wall 85
Fig. 3.30: Toothing 86
Fig. 3.31: Tools in masonry work 88
Fig. 3.32: Hollow concrete block 89
Fig. 3.33: Representation of composite masonry 90
Fig. 3.34: Single scaffolding 92
Fig. 3.35: Double scaffolding 92
Fig. 3.36 (a): Single frame cantilever 93
Fig. 3.36 (b): Double frame cantilever 93
Fig. 3.37: Suspended scaffolding 93
Fig. 3.38: Trestle scaffolding 94
Fig. 3.39: Dead shoring 96
Fig. 3.40: Raking shoring 97
Fig. 3.41: Flying shoring 98

4. Building Communication and Ventilation 107

Fig. 4.1: Component of doors 109
Fig. 4.2: Different types of single shutter panel door 110
Fig. 4.3: Details of a double shutter panel door with six panel 111
Fig. 4.4: Fully glazed door 112
Fig. 4.5: Details of the partly glazed and partly paneled door 112
Fig. 4.6: Details of the collapsible door 113
Fig. 4.7: Revolving door 113
Fig. 4.8: Rolling shutter door 114
Fig. 4.9: Solid/Laminated core flush door 116
Fig. 4.10: Hollow core flush door 117
Fig. 4.11: Components of the window 119
Fig. 4.12: Full paneled window 120
Fig. 4.13: Glazed window 121
Fig. 4.14: Partly paneled window 122
Fig. 4.15: Sliding window 123
Fig. 4.16: Louvered window 124
Fig. 4.17: Bay window 124
Fig. 4.18: Corner window 125
Fig. 4.19: Clerestory window 125
Fig. 4.20 (a): Gable window 126
Fig. 4.20 (b): Dormer window 126
Fig. 4.21: Hinges for doors and windows 128
Fig. 4.22: Handles 128
Fig. 4.23: Bolts 129

Fig. 4.24 (a): Door stopper 129
Fig. 4.24 (b): Pad lock 129
Fig. 4.24 (c): Cup board lock 129
Fig. 4.25: Ramp 131
Fig. 4.26: Lift or elevator 131
Fig. 4.27: Parts of staircase 132
Fig. 4.28 (a): Straight staircase 133
Fig. 4.28 (b): Plan of the straight staircase 133
Fig. 4.29 (a): Elevation of dog legged staircase 134
Fig. 4.29 (b): Plan of the dog legged staircase 134
Fig. 4.30: Open well staircase 135
Fig. 4.31 (a): Spiral staircase 135
Fig. 4.31 (b): Plan of the spiral staircase 135
Fig. 4.32 (a): Quarter-turn Staircase 136
Fig. 4.32 (b): Three-quarter-turn staircase 136
Fig. 4.33: Bifurcated staircase 136
Fig. 4.34: Stone staircase 137
Fig. 4.35: Brick staircase 138

5. Building Finishes 141
Fig. 5.1: Kota finish 142
Fig. 5.2: Granite floor finish 143
Fig. 5.3 (a): Ceramic tiles 144
Fig. 5.3 (b): Vitrified tiles 144
Fig. 5.4: Chequered tiles 145
Fig. 5.5: Paver Blocks 145
Fig. 5.6: Concrete flooring 146
Fig. 5.7: Alternate filling of panels 146
Fig. 5.8: Skirting tiles 147
Fig. 5.9: RCC slab 149
Fig. 5.10: Battened or mud roof 152
Fig. 5.11: King post truss and its components 153
Fig. 5.12: Queen post truss and its components 154
Fig. 5.13: Sponge finish 158
Fig. 5.14: Pebble finish 158
Fig. 5.15: Plasterboards 158
Fig. 5.16: Types of pointing 161
Fig. 5.17: Paint roller 163

LIST OF TABLES

Table 4.1: Size of Doors 118
Table 4.2: Size of Windows 127

CONTENTS

Foreword *iii*
Acknowledgement *v*
Preface *vii*
Outcome Based Education *ix*
Course Outcomes *x*
Guidelines for Teachers *xi*
Guidelines for Students *xii*
Abbreviations and Symbols *xiii*
List of Figures *xv*
List of Tables *xx*

1. Overview of Building Components........ 1-16
Unit Specifics 1
Rationale 1
Pre-requisites 1
Unit Outcomes 2
1.1 Introduction to Building Construction 2
1.2 Classification of Buildings (As per NBC) 2
1.2.1 Group A: Residential Buildings 3
1.2.2 Group B: Educational Buildings 3
1.2.3 Group C: Institutional Buildings 3
1.2.4 Group D: Assembly Buildings 4
1.2.5 Group E: Business Buildings 4
1.2.6 Group F: Mercantile Buildings 4
1.2.7 Group G: Industrial Buildings 5
1.2.8 Group H: Storage Buildings 5
1.2.9 Group I: Hazardous Buildings 5
1.3 Classification of Buildings (As per types of Construction) 5
1.3.1 Load-bearing Structures 5
1.3.2 Framed Structures 7
1.3.3 Composite Structures 7
1.4 Components of Buildings 7
1.4.1 Substructure 7
1.4.2 Superstructure 9
1.4.2.1 Walls 9
1.4.1.2 Cavity wall 9
1.4.2.3 Partition wall 10
1.4.2.4 Sill 10

1.4.2.5 Lintel 11
1.4.2.6 Doors and windows 11
1.4.2.7 Floor 11
1.4.2.8 Mezzanine floor 12
1.4.2.9 Roof 12
1.4.2.10 Structural Members of Building 13
1.4.2.11 Parapet 14
1.5 Functions of the building components 14
Unit Summary 14
Exercises 15
Know More 16
References and Suggested Readings 16

2. Construction of Substructure.. 17-55
Unit Specifics 17
Rationale 17
Pre-requisites 17
Unit Outcomes 18
2.1 Introduction 18
2.2 Job Layout 18
2.2.1 Factors Affecting Job Layout 19
2.2.2 Site Clearance 20
2.2.3 Centre Line Method of Job Layout 20
2.2.4 Face Line Method of Job Layout 21
2.2.5 Precautions 22
2.3 Earthwork 23
2.3.1 Excavation for Foundation 23
2.3.2 Timbering and Strutting 24
2.3.3 Earthwork for Embankment 26
2.3.4 Materials for Plinth 26
2.3.5 Tools and Plants for Excavation 27
2.4 Foundation 29
2.5 Functions of Foundation 29
2.6 Types of Foundation 30
2.7 Shallow Foundation 30
2.7.1 Depth of Shallow foundation 30
2.7.2 Spread footings 31
2.7.3 Combined footings 37
2.7.4 Strap footings 40
2.7.5 Raft foundations 41
2.8 Deep Foundation 41

2.8.1 Pile foundation 42
2.8.2 Pier foundation 44
2.8.3 Caissons 44
2.9 Pumping Methods for Dewatering 46
2.9.1 Well Points 47
2.9.2 Vacuum Well Points 49
2.9.3 Deep well system 49
2.10 Coffer DAM 50
Unit Summary 52
Exercises 53
Know More 55
References and Suggested Readings 55

3. Construction of Superstructure 56-106
Unit Specifics 56
Rationale 56
Pre-requisites 57
Unit Outcomes 57
3.1 Introduction 57
3.2 Stone Masonry 57
3.3 Terms used in Stone Masonry 58
3.4 Types of Stone Masonry 59
3.4.1 Rubble Masonry 59
3.4.1.1 Dry Rubble Masonry 59
3.4.1.2 Random Rubble Masonry 60
3.4.1.3 Square Rubble Masonry 61
3.4.1.4 Polygon Walling Masonry 62
3.4.1.5 Flint Walling Masonry 62
3.4.2 Ashlar Masonry 62
3.4.2.1 Ashlar Rough Tooled Masonry 63
3.4.2.2 Ashlar Fine Tooled Masonry 63
3.4.2.3 Ashlar Rock Faced Masonry 64
3.4.2.4 Ashlar Chamfered Masonry 64
3.4.2.5 Ashlar Coursed Masonry 64
3.4.2.6 Ashlar Facing Masonry 65
3.5 Joints in Stone Masonry and their Purpose 65
3.6 Selection of Stones 66
3.7 Precautions in the Construction of Stone Masonry 67
3.8 Brick Masonry and the Terms used in Brick Masonry 67
3.9 Bonds in Brick Masonry 70
3.9.1 Stretcher Bond 71

3.9.2 Header Bond 71
3.9.3 English Bond 72
3.9.4 Flemish Bond 74
3.10 Requirements of a Good Brick 77
3.11 Junctions in Brick Masonry and their Procedure 78
3.11.1 Tee-Junction 78
3.11.2 Intersection or cross-junction 83
3.11.3 Squint Junctions 84
3.12 Purpose of Junctions 86
3.13 Precautions in Construction of Brick Masonry 86
3.14 Comparison Between Stone and Brick Masonry 87
3.15 Tools and Plants for Stone and Brick Masonry Construction 88
3.16 Hollow Concrete Block Masonry 89
3.17 Composite Masonry 90
3.18 Scaffolding and its Purpose 91
3.19 Types of Scaffold 91
3.19.1 Single Scaffold 91
3.19.2 Double Scaffold 92
3.19.3 Cantilever Scaffolding 93
3.19.4 Suspended Scaffolding 93
3.19.5 Trestle Scaffolding 94
3.19.6 Steel Scaffolding 94
3.20 Process of Erection of Scaffolding 94
3.21 Dismantling of Scaffolding 95
3.22 Shoring and its Purpose 95
3.23 Types of Shoring 96
3.23.1 Dead Shoring 96
3.23.2 Raking Shoring 96
3.23.3 Flying Shoring 97
3.24 Underpinning 98
3.25 Definition of Formwork 99
3.26 Requirements of Formwork 99
3.27 Materials used in Formwork 100
3.28 Types Of Formwork 100
3.28.1 Timber Formwork 100
3.28.2 Plywood Formwork 101
3.28.3 Steel Formwork 101
3.28.4 Plastic Formwork 101
3.28.5 Aluminium Formwork 102
3.29 Removal of Formwork 102

Unit Summary 102
Exercises 104
Know More 105
References and Suggested Readings 106

4. Building Communication and Ventilation 107-140
Unit Specifics 107
Rationale 107
Pre-requisites 107
Unit Outcomes 107
4.1 Doors 108
4.2 Components of the Door 108
4.3 Paneled Doors 110
4.4 Glazed Doors 112
4.5 Partly Glazed/ Partly Paneled Doors 112
4.6 Collapsible Doors 112
4.7 Revolving Doors 113
4.8 Rolling Shutter Doors 114
4.9 Flush Doors 115
4.10 Size of Doors (As per bis Norms) 117
4.11 Windows 118
4.12 Component of Windows 119
4.13 Types of Windows 120
4.13.1 Full Paneled Window 120
4.13.2 Glazed Window 121
4.13.3 Partly Paneled Window 122
4.13.4 Wooden Windows 122
4.13.5 Steel Windows 122
4.13.6 Aluminium Windows 123
4.13.7 Sliding Windows 123
4.13.8 Louvered Windows 124
4.13.9 Bay Window 124
4.13.10 Corner Window 125
4.13.11 Clerestory Window 125
4.13.12 Gable and Dormer Window 126
4.13.13 Skylight Window 126
4.14 Sizes of Windows (as per Bis Norms)
4.15 Ventilators 127
4.16 Fixures and Fastenings for the Doors and Windows 128
4.17 Functions and Materials of Window Sill and Lintels 130
4.18 Shed/Chhajja 130

4.19 Means of Vertical Communication 131
4.20 Terms used in Staircase 132
4.21 Types of Staircases (on the Basis of Shapes) 133
4.21.1 Straight Staircase 133
4.21.2 Dog-legged Staircase/ Half-turn Staircase 134
4.21.3 Open Well Staircase 135
4.21.4 Spiral Staircase 135
4.21.5 Quarter-turn and Three Quarter-turn 135
4.21.6 Bifurcated Staircase 136
4.22 Types of Staircases (on the Basis of Materials) 137
4.22.1 Stone Staircase 137
4.22.2 Brick Staircase 138
4.22.3 RCC Staircase 138
4.22.4 Wooden and Metal Staircase 138
Unit Summary 138
Exercises 139
Know More 140
References and Suggested Readings 140

5. Building Finishes 141-166
Unit Specifics 141
Rationale 141
Pre-requisites 141
Unit Outcomes 141
5.1 Floors And Roofs 142
5.2 Types of Floor Finish and its Suitability 142
5.2.1 Kota Floor Finish 142
5.2.2 Marble Floor Finish 143
5.2.3 Granite Floor Finish 143
5.2.4 Ceramic Tiles and Vitrified Tiles 144
5.2.5 Chequered Tiles 145
5.2.6 Paver Blocks 145
5.2.7 Concrete Flooring 145
5.2.8 Wooden Flooring 147
5.2.9 Skirting and Dado 147
5.3 Process of Laying and Construction 148
5.4 Finishing and Polishing of Floors 148
5.5 Roofing Materials 148
5.5.1 Reinforced Cement Concrete (RCC) 149
5.5.2 Mangalore Tiles 150
5.5.3 Asbestos Cement (AC) Sheet 150

5.5.4 Galvanized Iron (GI) Sheet 150
5.5.5 Plastic and Fiber Sheet 151
5.6 Terms Used In Roofs 151
5.7 Types Of Roofs 152
5.7.1 Flat Roof 152
5.7.2 Pitched Roof 153
5.7.2.1 King Post Truss 153
5.7.2.2 Queen Post Truss 154
5.8 Plastering and its Necessity 155
5.9 Procedure of Plastering 155
5.10 Double Coat Plastering 156
5.11 Single Coat Plastering 156
5.12 Rough Finish and Neeru Finish 156
5.13 Plaster of Paris (POP) 157
5.14 Special Plasters 157
5.15 Stucco Plaster 157
5.16 Sponge Finish 158
5.17 Pebble Finish 158
5.18 Plasterboard 158
5.19 Wall Claddings 159
5.20 Precautions to be Taken in Plastering 159
5.21 Defects In Plastering 160
5.22 Pointing 160
5.22.1 Necessity of Pointing 160
5.22.2 Types of Pointing 161
5.23 Painting and its Necessity 162
5.24 Surface Preparation for Painting 162
5.25 Methods of Application 162
Unit Summary 163
Exercises 164
Know More 166
References and Suggested Readings 166

Appendices **167-176**

References for Further Learning **177**

CO and PO Attainment Table **178**

Index **179-180**

1 Overview of Building Components

UNIT SPECIFICS

Through this unit we have discussed the following aspects:

- Classification of the buildings according to the norms of the National Building Code of India (SP 7: 2016)
- Functions of the components of the building
- Overview of the sub-structure and its components
- Overview of the superstructure and its components

For helping the students in the visualization of the content and to enhance their practical knowledge, the figures are provided in 2 D and 3D views.

This unit contains short and long answer-type questions along with multiple choice questions, a list of references, and suggested reading is given so that one can go through them for practice. Some QR codes are given which can be scanned for more information on various topics of interest and to get extra knowledge.

RATIONALE

This unit on overview of building components helps the students to develop an idea about the classification of the buildings according to the National Building Code of India. It explains the various components of the buildings. The functions of the components of the buildings are discussed. A brief discussion about the substructure and superstructure of the building is provided. The content covered in this unit is important to build a foundation for studying the subject further.

Building construction is an important subject for civil engineers and building planners. Knowledge of the building components, techniques of construction, the layout of buildings, etc. is essential for students of civil engineering. The students can apply the knowledge gained from this subject in the field of quantity surveying, building construction supervision, material testing, engineering drawing, and building design.

PRE-REQUISITES

No prerequisites are required for studying this unit.

UNIT OUTCOMES

The list of outcomes of this unit is as follows:

U1-O1: Identify different types of buildings based on the National building code of India

U1-O2: Differentiate the types of the building based on construction

U1-O3: Explain different components of the buildings

U1-O4: Realize the role of different components of the building

Unit-1 Outcome	EXPECTED MAPPING WITH COURSE OUTCOMES (1 - Weak Correlation; 2 - Medium Correlation; 3 - Strong Correlation)				
	CO-1	CO-2	CO-3	CO-4	CO-5
U1-O1	1	1	1	-	1
U1-O2	1	2	1	1	1
U1-O3	3	1	2	-	-
U1-O4	2	-	-	-	-

1.1 INTRODUCTION TO BUILDING CONSTRUCTION

Shelter is the most important need the human beings after food and clothes. Even the birds, animals, and other creatures also need shelter for survival. The buildings in a nation represent its level of development. Human shelters have changed significantly with time. Humans have reached from the caves to these comfortable homes with the advancement of time. A good shelter gives confidence and reason to live with dignity to human beings.

1.2 CLASSIFICATION OF BUILDINGS (AS PER NBC)

You would have observed many types of buildings in your vicinity. Some are used for storing goods and others are used as shops. You are also living in a structure called home. Have you ever thought, how many types of buildings are there? According to the National Building Code (NBC) of India, a building is classified into nine categories depending on its usage. The buildings as per NBC criteria are classified into the following categories:

- **A.** Residential Buildings
- **B.** Educational Buildings
- **C.** Institutional Buildings
- **D.** Assembly Buildings
- **E.** Business Buildings
- **F.** Mercantile Buildings
- **G.** Industrial Buildings
- **H.** Storage Buildings
- **I.** Hazardous Buildings

SCAN ME

1.2.1 Group A: Residential Buildings

The buildings which have sleeping arrangements with or without cooking and dining facilities are known as residential buildings. According to NBC, Residential buildings are further divided into five sub-categories as follows:

- **A-1: Lodging and rooming houses:** This type of building consists of a sleeping arrangement for fewer than 40 persons and doesn't have any cooking facility. The building or group of buildings with 40 beds or fewer lies in this category.
- **A-2: One or two-family private dwelling:** The buildings which provide accommodation to one or two families with a bed capacity of 20 or less are known as A-2 type of residential buildings.
- **A-3: Dormitories:** In dormitories, a group of sleeping accommodations is provided for public use. The users in the case of A-3 type residential buildings don't belong to the same family.
- **A-4: Apartment houses:** Apartment houses are also known as flats. The buildings which have three or more quarters for living independently are known as apartments or flats. The cooking and dining facilities are provided in such types of buildings.
- **A-5: Hotels:** The buildings which provide sleeping accommodations with or without dining facilities are classified as hotels.

1.2.2 Group B: Educational Buildings

The buildings which are used for providing education or training are known as educational buildings. Schools, colleges, and universities come under the category of education buildings. Group B buildings are further divided into two categories.

- **B-1:** The schools up to the senior secondary class system come under the B-1 category of educational institutions
- **B-2:** The teaching and training institutions that do not belong to the B-1 category are categorized under the B-2 category.

1.2.3 Group C: Institutional Buildings

The buildings which are used for care, treatment, imprisonment, etc. are known as institutional buildings. The institutional buildings consist of sleeping arrangements. Group C buildings are further divided into the following three sub-categories:

- **C-1: Hospital and Sanatoria:** The buildings which are used for accommodating people who are suffering physically. Hospitals, nursing homes, healthcare centers, etc. come in this category.
- **C-2: Custodial Institutions:** The buildings which are used for the protective care of children, old age people, teenagers, etc. are known as custodial institutions. The orphanage, old age homes, juvenile homes, etc. lie under the C-2 category of institutional buildings.
- **C-3: Penal and Mental Institutions:** The buildings which are used for accommodating persons for restriction of their liberty are called penal and mental institutions. C-3 type institutional buildings include mental asylums, prisons, jails, etc.

1.2.4 Group D: Assembly Buildings

The buildings which are used for accommodating the gathering of 50 or more people are known as assembly buildings. The people can gather for social, religious, recreational purposes, etc. According to NBC, the assembly buildings are further classified into the following seven categories:

D-1: The building which has a stage and fixed seats with an accommodating capacity of more than 1000 viewers are classified under the D-1 category of assembly buildings. Theatres, exhibition rooms, cinema halls, auditoriums, etc. are the D-1 category of assembly buildings.

D-2: The building which has a stage and fixed seats with an accommodating capacity of fewer than 1000 viewers are classified as a D-2 category assembly building.

D-3: The buildings which neither have a permanent stage nor fixed seats but have an accommodating capacity of 300 persons or more are known as D-3 assembly buildings.

D-4: The buildings which have the same characteristics as D-3 buildings but an accommodating capacity is less than 300, are known as D-4 assembly buildings.

D-5: All the other structures for the gathering of people which are not covered in the D-1 to D-4 sub-categories lie in D-5 assembly buildings.

D-6: The buildings which have mixed facilities of shopping, theatre, gaming, etc. lie under the D-6 category of assembly buildings. Modern-day shopping malls and multiplexes are examples of D-6 assembly buildings.

D-7: Any underground or elevated structure for the gathering of people for different purposes which is not covered by D-1 to D-6 categories is kept under the D-7 assembly building category.

1.2.5 Group E: Business Buildings

The buildings which are used for business transactions, and for maintaining the records and accounts are known as business buildings. Libraries, banks, offices, laboratories, etc. lie in this category of buildings. Group E buildings are further divided into three categories:

E-1: The buildings which are used as offices, banks, and other establishments are categorized as E-1 type of business buildings.

E-2: The buildings which are used as laboratories, research laboratories, testing houses, etc. are categorized as E-2 business buildings.

E-3: The buildings where groups of computers are installed are known as E-3 business buildings.

E-4: Telephone exchanges are categorized in this category.

E-5: Television stations and radio broadcasting stations lie in this category.

1.2.6 Group F: Mercantile Buildings

The buildings which are used as shops, retail stores, markets, wholesale shops, etc. are known as mercantile buildings. Mercantile buildings are further divided into the following three categories:

F-1: The shops or stores having an area equal to or less than 500 m^2 are known as F-1 mercantile buildings.

F-2: The shops or stores with an area of more than 500 m^2 are known as F-2 mercantile buildings.

F-3: The underground structures which are used as shops or stores are categorized as F-3 mercantile buildings.

1.2.7 Group G: Industrial Buildings

The buildings in which manufacturing, fabricating, assembling, and processing of materials and products are performed are known as industrial buildings. The buildings used by various industries like dairies, tanneries, dry cleaning plants, pump stations, etc. the examples of group G buildings. Industrial buildings are further classified into the following three categories:

G-1: The buildings which are used for low hazard industry come in this category. The materials and products accommodated in the G-1 type of building comparatively have low combustibility and there are very less chances of self-propagating fires.

G-2: The buildings which are used for moderate hazard industry come in this category. The material and products accommodated in such type of building burn at a medium speed but toxic fumes or explosions during a fire are not generated.

G-3: The buildings which are used for high hazard industry come in this category. The materials and products accommodated in such type of building burn rapidly. Toxic fumes and explosions during the fire can be generated.

1.2.8 Group H: Storage Buildings

The building or part of the building that is used to store the goods, products, merchandise, etc. except the hazardous material is known as a storage building. Storehouses, grain storage, fruit stores, garages, cold storage, animal shelters, etc. are examples of group H buildings.

1.2.9 Group I: Hazardous Buildings

The buildings which are used to store and process extremely inflammable material or goods are termed as hazardous buildings. These hazardous materials can produce excessive toxic fumes or explosions during a fire. Group I buildings may store the explosives or chemicals which are to be kept with great care at specific pressure or temperature. Acetylene, rocket fuel, liquefied petroleum gas (LPG), compressed natural gas (CNG), etc. are accommodated in hazardous buildings.

1.3 CLASSIFICATION OF BUILDINGS (AS PER TYPES OF CONSTRUCTION)

A building has many components like walls, floor, roof, etc. The load is transferred to the foundation from the components of the building. The buildings can be classified into the following three categories based on the construction:

1.3.1 Load Bearing Structures
1.3.2 Framed Structures
1.3.3 Composite Structures

1.3.1 Load-bearing Structures

The structure in which a load of the roof is transferred through the walls to the foundation. The load-bearing structures are used for the construction of small buildings. Small structures like temples, small houses, rural buildings, and low-rise buildings are examples of load-bearing structures.

Advantages

The major advantages of load-bearing structures are as follows:

a. If the soil conditions are favorable, the load-bearing structures prove to be very cost-effective.
b. The masonry used in load-bearing structures offers good fire resistance characteristics.
c. The load-bearing structure has good sound insulation properties.

Disadvantages

The major disadvantages of load-bearing structures are as follows:

a. The load-bearing structures are suitable only for up to two to three-floor buildings.
b. These types of structures are not earthquake resistant and less preferred than framed structures.
c. Long spans are not possible for load-bearing structures.
d. This approach reduces the floor area of the building.

Figure 1.1 shows the load-bearing structure.

Fig. 1.1: Load-bearing structure

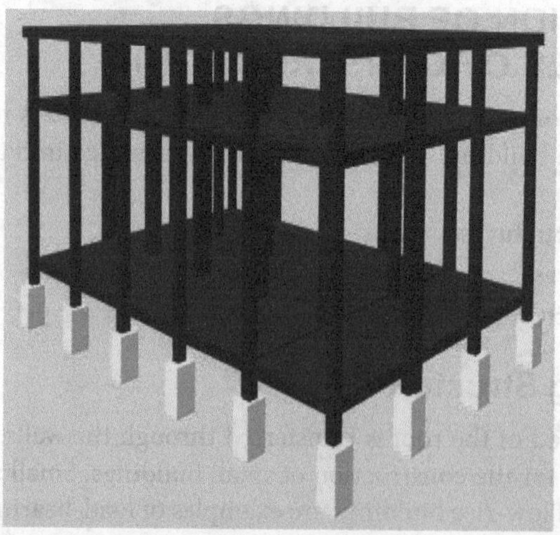

Fig. 1.2: Framed structure

1.3.2 Framed Structures

The building components like beams, columns, and slabs are integrated to form a framed structure. A framed structure consists of a combination of these building components. This process of combining the structural components is known as framing.

In the framed structure, the load is transferred through this frame to the foundation. Figure 1.2 shows the framed structure. The frames can be made of the materials like concrete, timber, and steel. Concrete frames are used commonly for framed building structures.

Concrete structures are very durable and cost-effective. The wooden frames are light in weight and flexible. They are good insulators of electricity and sound. The only problem with wooden frames is their durability. The wooden frames are prone to insect attack. Steel frames provide good resistance to fire. The steel frames are also light in weight and provide higher tensile strength.

The framed structures offer the following advantages:

a. The framed structure offers good stability and strength. These structures can easily withstand earthquake loads and wind loads.
b. These structures can be used to construct multi-story buildings.
c. These structures can absorb the shocks and vibrations in a better way as compared to load-bearing structures.

1.3.3 Composite Structures

The load-bearing and framed structures are combined to form the composite structures. The composite structures are used for long spans. The advantages of load-bearing structures and framed structures are coined in composite structures. The walls in the composite buildings act as load-bearing structures while the columns run through the structure as framed structures. Some examples of composite structures are long shed diaries, storage houses, workshops, etc.

1.4 COMPONENTS OF BUILDINGS

The building components are the structures that serve the purpose of load transfer, supporting, and protecting the building. A building can be divided into the following two parts:

a. **Substructure:** Substructures or foundations can be defined as all lower structures of building below ground level which helps in transmitting loads of superstructures to the soil. A foundation is that part of the building structure which is in direct contact with the supporting soil and transfers structural loads.
b. **Superstructure:** The superstructure can be defined as all the structures that lie above the ground level both internally and externally.

1.4.1 Substructure

A substructure transfers the load from the various building components to the earth. The substructure remains beneath the ground. Figure 1.3 shows the substructure and its components. The substructure of a building consists of the following two parts:

a. **Foundation:** The basic function of the foundation is to prevent the building from collapsing by transferring the dead load, live load, wind load, earthquake load, and any other type of load to the subsoil in such a manner that it doesn't cause any settlement or shear failure of soil. Foundation should have sufficient bearing capacity and stability. It is therefore designed very carefully as the failure of the foundation cannot be noticeable till it affects the building. A detailed discussion of the foundation types and their functions is explained in Unit II.

b. **Plinth:** The part of the structure between the surface of the surrounding subsoil and the surface of the raised floor of the building, immediately above the ground is known as a plinth. The width of the plinth beam should be equal to the width of the foundation at the ground level. It should have a minimum depth of 20 cm and the concrete used for construction should have a minimum of 20 MPa strength. Some of the important advantages of the plinth in building construction are as follows:
   - ➢ The plinth distributes loads of the superstructure evenly to the foundation.
   - ➢ It acts as a barrier to the dampness and moisture reaching the superstructure.
   - ➢ The plinth protects from water seepage in a framed structure.
   - ➢ It protects the building from cracks at the time of settlement of the foundation.
   - ➢ It provides durability and stability to the superstructure.
   - ➢ It gives a better aesthetic appeal to the structure.

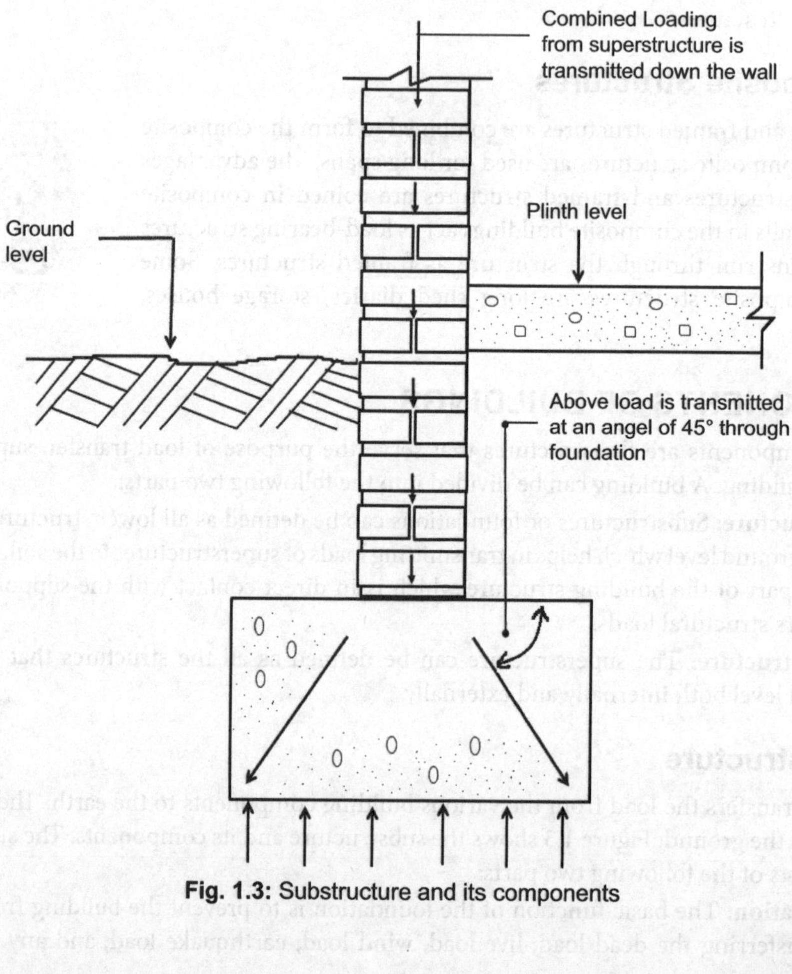

Fig. 1.3: Substructure and its components

1.4.2 Superstructure

The superstructure is the visible part of a building. The components of the superstructure are visible and you can see them at your home too. Doors, windows, floor, roof etc. are the parts of the superstructure. The various components of the superstructure are discussed in the subsequent sections. Figure 1.4 shows the different components of the superstructure.

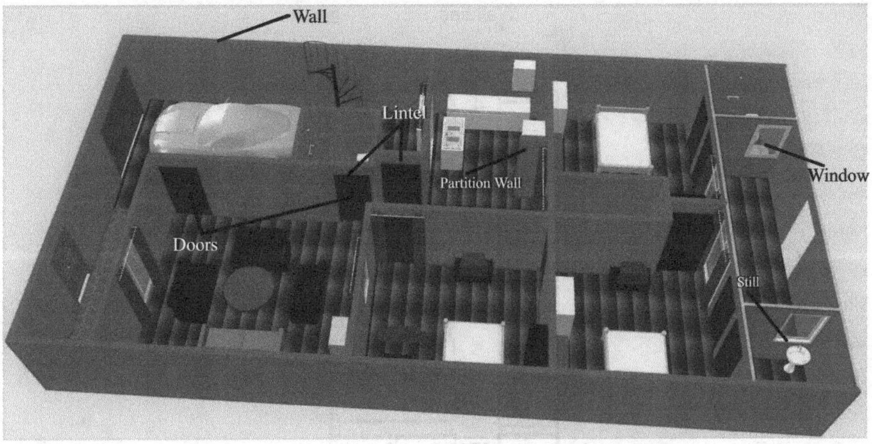

Fig. 1.4: Components of the superstructure

1.4.2.1 Walls

Walls are the most important parts of the building's superstructure. A wall can be used for the transfer of load or for just partition purposes. The walls which are used to transfer the loads are known as load-bearing walls while the walls which don't act as a load-bearing structure are known as non-load-bearing walls. Some of the salient features of a wall in building construction are given as follows:

- It should be strong to carry its own weight.
- It must be capable of carrying and transmitting the load of the roof to the substructure.
- It must be strong enough to support the opening of doors and windows.
- It helps in the space management of buildings to make them more useful.
- It protects the occupants from heat, rain, and cold and provides privacy.
- It acts as a sound barrier.
- It should be fire-resistant.

Walls are of different types depending on their usage and location such as cavity walls, partition walls, retaining walls, compartment walls, dwarf walls, parapet walls, and curtain walls. The cavity wall and partition wall are discussed in the next section. Figure 1.4 shows the walls.

1.4.1.2 Cavity wall

Cavity walls are types of walls made up of two different walls joined with the help of wall ties. In this type of wall, an outer wall is attached with the help of ties to the inner wall separated by cavity/airspace. Metal strips are used as wall ties. Generally, wall ties are placed at the gap of 90 cm along the length and 45 cm in the direction of height. The thickness of the external wall is kept at 10 cm and the thickness of the inner wall is as per the imposed load. The cavity wall protects the building from external wall moisture and rainwater and also provides space for the installation of thermal insulation. Buildings with cavity walls face a lesser impact of external temperature. This type of building remains cooler in summer and warm in winter. As the inner wall is protected from the external atmosphere, the life of the building also increases. The cavity starts from the floor level to the roof of the building and is covered from the top to prevent the penetration of rainwater. Figure 1.5 shows the sketch of a cavity wall.

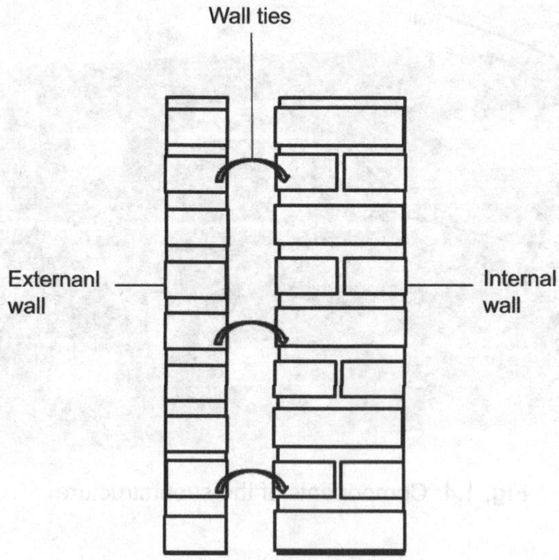

Fig. 1.5: Cavity wall

1.4.2.3 Partition wall

These types of walls are generally non load-bearing walls that are used to separate the internal common space of the building. This type of wall only carries its own weight. So, no foundation is required as in the case of a normal wall. It can be raised directly from the floor level. If the partition wall is not constructed up to the height of the roof and has a height of around 5 feet to 6 feet, then it is called a curtain wall. The partition walls can be easily seen in banks, railway stations, offices, hospitals, hotels etc. Materials used for the construction of partition walls are brick, timber, plywood, glass, aluminum sheets, PVC board, etc.

In modern-day buildings, where the construction is carried out as framed structures, the utility of partition wall has increased. These types of walls provide flexibility as wall positions can be easily reconfigured and inexpensive without impacting the structure of the building. They can be easily designed and constructed with openings for doors, windows, sockets, etc. The partition wall specification depends on the availability of materials, cost, ease of construction, and finishing. Figure 1.4 shows the partition wall.

1.4.2.4 Sill

Sill a horizontal surface made up of mortar located below the window and having a width equal to the width of the wall below the window. It provides a smooth finished surface and support to the window. Materials used for the construction of sill are reinforced cement concrete, plain cement concrete, wood, steel, etc. Sill helps in preventing the moisture and passage of water to reach up to the window.

1.4.2.5 *Lintel*

Lintels are horizontal stripes such as beams or any structural member, which are used to cover the opening and support the structure above the opening of the doors and windows. Materials used for constructing lintel are wood plank, bricks, reinforced cement concrete, stone, and rolled steel section infused in cement concrete.

The width of the lintel is provided equally to the width of the building wall. The vertical load above the opening of the door and window is carried by the lintel and transmitted to the side wall from its end. Figure 1.4 shows the lintel.

1.4.2.6 *Doors and windows*

A door is a moving structure within the wall of the building used to allow the entrance or block the passage of the building or a room. The door is designed in such a way that any person can enter without any complications. They are generally located at the corner of the room in such a way that they cover the minimum floor area of the room while opening and provide security to the building or room on closing.

The door is attached to the wall with the help of a frame and shutter. The wood used for the construction of the door should conform to IS 12896 (1990). Commonly used woods for the door are teak, cedar, kail, etc. Nowadays, different materials are available in the market for the construction of the doors like aluminum, PVC, glass, soft steel, etc.

Window is installed in the building to properly flow light and ventilation. If there is more than one window in the room, then wherever possible windows are kept opposite to each other for better circulation of air. The window shutter in the external wall of the building should open outside. The sill of the window is kept at 75 to 100 cm above the floor level for better natural light. A detailed discussion about doors and windows is given in Unit-IV. Figure 1.4 shows the doors and windows in a building.

1.4.2.7 *Floor*

A floor may be defined as the vertical partition of a building one over the other to increase the accommodating capacity of the building in a limited area. A floor is a flat and smooth surface that is designed to support the occupants, furniture, equipment, etc. the floor at the ground is called 'ground floor' and the floor at the first storey is called 'first floor'. A floor has the following two components:

a. **Sub-floor:** A sub-floor provides sufficient support to the floor covering and the load on the floor covering. A sub-floor imparts the necessary strength to bear the superimposed load.

b. **Floor covering:** Floor covering is a flat, even, smooth and durable surface that provides the face of the floor. The floor covering is impervious in nature and aesthetically pleasing.

The floors are classified as ground floor and upper floors. In the ground floor, the floor rests on the ground while in the case of upper floors the floors are supported by some other means like slab action, timber framing, etc. Figure 1.6 shows the system of floors in a building.

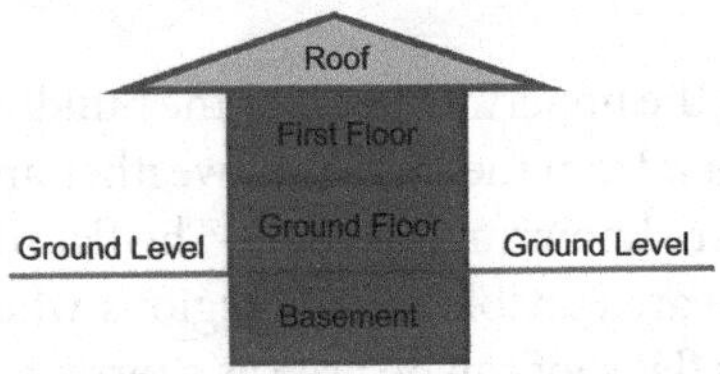

Fig. 1.6: Floors of a house

The ground floors can be further divided into three categories which are suspended floors, solid floors, and basement floors. The suspended floors do not rest on the ground and have blank space underneath. The solid floor rests on the ground and can take more load compared to the suspended floor. The floor covering rests on a floor of solid concrete. The solid ground floor has more strength and shows less movement as compared to the other floors. The basement ground floor is provided where the underground structure is provided in the building.

1.4.2.8 Mezzanine floor

The mezzanine floor may be defined as an intermediate floor that lies between the floor and the ceiling. Mezzanine floors are temporary or semi-permanent structures that are used to maximize the use of the space in the building. These floors only extend to some portion and are not counted as main floors. The minimum size of the mezzanine floor is to be 9.5 m^2. The height of the mezzanine floor should be from 2.20 m to 2.70 m. Figure 1.7 shows the image of the mezzanine floor.

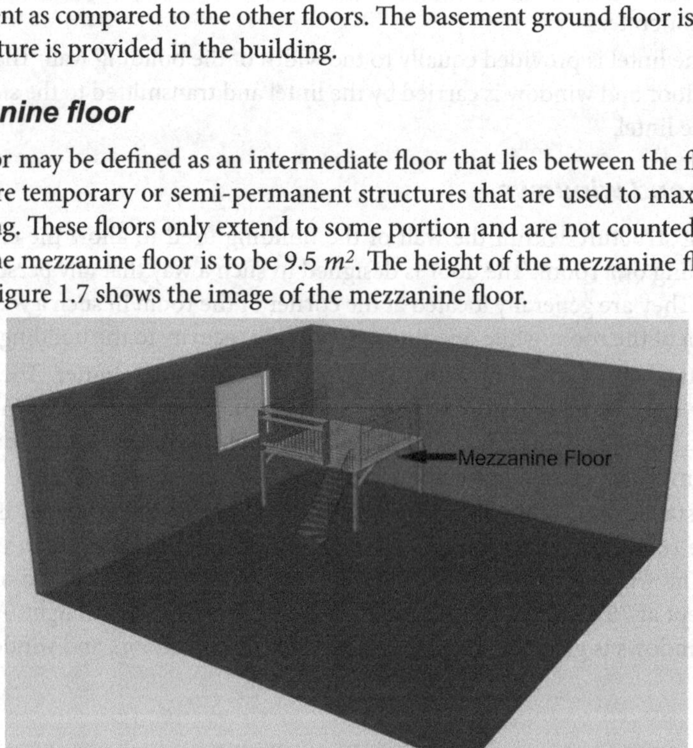

Fig. 1.7: Mezzanine Floor

1.4.2.9 Roof

A roof may be defined as the uppermost part of the building that acts as a structural covering to the building and protects it from the impact of weather and external impacts. The roofs are constructed as the floors but the shapes remain different. The roofs can be of the following three types:

a. Flat roof
b. Sloping roof
c. Curved roof

a. Flat roof

A roof may be defined as the uppermost part of the building that acts as a structural covering to the building and protects it from the impact of weather and external impacts. The roofs which have a very gentle slope are known as flat roofs. The flat roofs have a slope gradient of less than 10°. These types of roofs are suitable for the regions where the rainfall quantity remains less throughout the year. The flat roof can be used as a terrace. It is necessary to provide some slope in the flat roofs for the proper drainage during the rainfall season. The flat roofs require less material for construction. These types of roofs are stable against the action of winds. In multi-storey buildings, only flat roofs can be constructed as it is easy to construct the upper floor on flat roofs.

b. **Sloping roof**

If the slope of the roof is more than 10°, it is known as the sloping roof. The sloping roofs are more suitable for regions where the quantity of rainfall is high. These roofs are cost-effective and can be provided for long spans. The regions where snowfall is a common phenomenon, sloping roofs are mandatory. The snow is easily drained out from the sloping roofs.

c. **Curved roofs**

Curved roofs are relatively new types of roofs. These types of roofs are relatively durable and look aesthetically pleasing. The curved roofs have very good drainage properties. These types of roofs are more suitable for the storage buildings as these increase the accommodating capacity of the storage buildings. The curved roofs are designed by specialists and require skilled labour. These types of roofs require less maintenance. The curved roofs require very sophisticated design capabilities and the cost may be higher due to the specific design requirements.

1.4.2.10 Structural Members of Building

The structural members of the building provide stability and strength to the building. The load from the building is transferred to the foundation through the structural members of the building. There are three structural members in a typical building: slabs, beams, and columns. Figure 1.8 shows the structural members of the building.

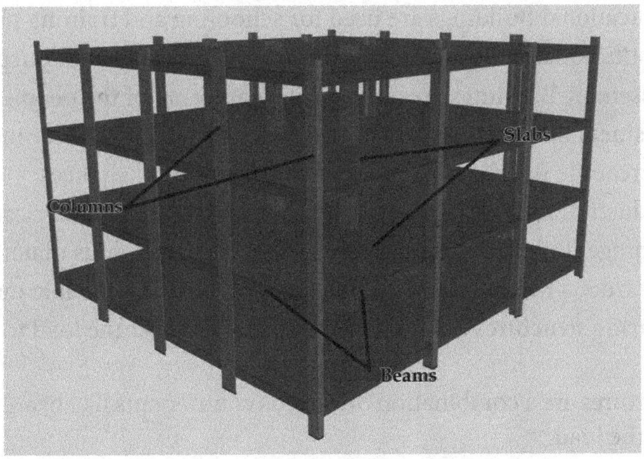

Fig. 1.8: Structural Members of the Building

- **Slabs** are the flat horizontal members which take the load of the people, goods, equipment, etc., and transfer it to the beams. The slabs are used to make flat surfaces like floors and roofs. The slabs are made of concrete generally and can be prefabricated too.
- **Beams** are the horizontal members which take the load from the slabs and transfer it to the columns. The beams take the vertical loads.
- **Columns** are the vertical members which take the load from the beams and transfer it to the foundation. The column is a compression member of the building as it counteracts the compressive load from the building components.

1.4.2.11 Parapet

A parapet wall may be defined as a wall that is constructed along the perimeter of the roof. These are short in height. The parapet walls are used to provide safety to roof users. The parapet wall helps in the prevention of accidents. It is aesthetically pleasing and provides the users with privacy. A parapet wall can be constructed of masonry, steel, stones, etc. The parapet walls can be provided in the form of railing too.

1.5 FUNCTIONS OF THE BUILDING COMPONENTS

The building components must fulfill the following conditions:

(a) They should satisfy the function for which they were constructed. For example, load-bearing components should transfer the load to the other members.
(b) They must be stable to resist the loads coming on them.
(c) They must be durable to serve their function and continue till their life.
(d) They must be economical and can be achieved with minimum resources available.

UNIT SUMMARY

- According to the National Building Code of India, 2016, Part IV, buildings can be classified into nine categories from Group A to Group I
  - Group A: Residential Buildings are used for sleeping and accommodation purposes.
  - Group B: Educational Buildings are used for schooling and training purposes.
  - Group C: Institutional Buildings are used for care and restriction purposes.
  - Group D: Assembly Buildings are used for the gathering of the people.
  - Group E: Business Buildings are used for the transactions of the business.
  - Group F: Mercantile Buildings are used as shops and marketplaces.
  - Group G: Industrial Buildings are used for the production and fabrication of products.
  - Group H: Storage Buildings are used for storing non-hazardous material and goods.
  - Group I: Hazardous Buildings are used to store highly inflammable materials.
- The load-bearing structures are the structures that transfer the load to the foundation through the walls.
- Framed structures are a combination of various components like beams, columns, and slabs for transferring the load.
- A building has two components which are substructure and superstructure.
- A substructure consists of the invisible part of the building. Foundation and plinth are the components of the substructure.
- The superstructure consists of the visible components in the building like doors, windows, sill, floors, etc.
- Slabs, beams, and columns are the structural members of the building which provide strength and stability to the structure.

EXERCISES

Multiple Choice Questions

1. According to NBC, industrial buildings lie in which group of buildings?
   (a) Group A (b) Group G
   (c) Group C (d) Group I
2. Which type of building come in Group C?
   (a) Institutional (b) Residential
   (c) Assembly (d) Industrial
3. Which one of the following is not part of the building?
   (a) Substructure (b) Semi-structure
   (c) Superstructure (d) All are the part of the buildings
4. Single-storey houses made of brick masonry are generally:
   (a) Framed structure (b) Load-bearing structure
   (c) Composite Structure (d) None
5. Black soil is a highly sensitive soil that changes its volume with moisture. As a civil engineer, for a three-floor building, what kind of structure will you suggest?
   (a) Load-bearing (b) Framed
   (c) Composite (d) None of the above
6. Which one of the following is true about a cavity wall?
   (a) It saves the inner structure from moisture (b) It keeps the rooms cool during summer
   (c) It increases the life of the inner structure (d) All of the above
7. The non-load-bearing wall that is constructed for the separation of portions in a house is known as:
   (a) Partition wall (b) Curtain wall
   (c) a and b both (d) None of the above
8. What is the minimum strength of concrete used for the construction of the plinth?
   (a) 10 MPa (b) 15 MPa
   (c) 20 MPa (d) 30 MPa
9. What is the height of the sill above the ground level?
   (a) 50 cm to 75 cm (b) 75 cm to 100 cm
   (c) 100 cm to 125 cm (d) More than 125 cm
10. Which one of the following is not the function of the parapet wall?
    (a) Safety (b) Aesthetics
    (c) Privacy (d) Strength to building

Answers of Multiple Choice Questions

1. (b), 2. (a), 3. (b), 4. (b), 5. (b), 6. (d), 7. (c), 8. (c), 9. (b), 10. (d)

Short and Long Answer Type Questions

1. Explain the classification of building according to the national building code.
2. Write short note on (a) Plinth, (b) lintel, (c) Sill
3. Explain the difference between (a) curtain wall and partition wall, (b) load bearing wall and Framed wall
4. Explain different components of the building.
5. What is a substructure? Explain with a neat sketch.
6. Explain the functions of the building components.
7. Why is a mezzanine floor provided in a building? What are the different functions of a mezzanine floor?
8. Explain the different functions of a parapet.

KNOW MORE

The curved roofs are introduced in the year 1920. The people who used to store the grains and grass for the animals wanted to increase the accommodating capacity of their storage building. So, for this purpose, curved roofs are introduced. At present, curved roofs have become a symbol of aesthetics and are used in various famous structures worldwide. To see some famous structures with curved roofs, you can scan the QR code given on the side of this text.

SCAN ME

REFERENCES AND SUGGESTED READINGS

1. Chudley, R., Greeno, R. *Building Construction Handbook*. Elsevier, Butterworth-Heinemann, Amsterdam, 2007.
2. Punmia, B. C., et al. *Building Construction: An Elementary As Well As Advanced for Engineering Students*. Laxmi Publications, 2016.
3. Iano, J., Allen, E. *Fundamentals of Building Construction: Materials and Methods*. Wiley. 2019.
4. *National Building Code of India (Part IV), Fire and Life Safety, 2016.*
5. "Load Bearing Structure and Components vs. Framed Structural System." The Constructor, 20 Sept. 2018, https://theconstructor.org/building/load-bearing-structure-components/24879/
6. Puybaret, E. *Sweet Home 3 D*, eTeks, 2016 (Open Access Software).

2 Construction of Substructure

UNIT SPECIFICS

Through this unit we have discussed the following aspects:

- To understand the layout of load-bearing structure and framed structure
- Process of excavation of foundation
- Overview of tools and plants used for earthwork
- Overview of functions and types of foundation
- Discussion on dewatering techniques
- Introduction of cofferdam

For helping the students in the visualization of the content and to enhance their practical knowledge, the figures are provided in 2D and 3D. The content in this unit provides practical knowledge and will help the students in decision-making during the fieldwork.

This unit contains short and long answer-type questions along with multiple choice questions, a list of references, and suggested reading is given so that one can go through them for practice. Some QR codes are given which can be scanned for more information on various topics of interest and to get extra knowledge. Some interesting facts are given under 'Know more'.

RATIONALE

This unit on the construction of substructure helps the students to develop an idea about the layout of the structures. It explains the process of the excavation of the foundation. The various tools and plants used for the earthwork are discussed. A brief discussion of the foundations and their types is provided. The various methods of dewatering are also discussed and an introduction about the cofferdams is given. The content covered in this unit is important to understand the process of construction of the substructure.

The knowledge of layout, excavation techniques, and construction of the foundation is essential for civil engineering students. The students can apply the knowledge of the construction of substructures during their fieldwork. The knowledge gained in this unit will help the students to identify the suitable foundation for different field conditions. The students can apply their knowledge in the selection and implementation of suitable dewatering techniques during the construction work. The concepts in this unit will help the student in effective supervision of construction works of the substructure.

PRE-REQUISITES

A basic knowledge of building components is required.

UNIT OUTCOMES

The list of outcomes of this unit is as follows:

U2-O1: To explain the process of the layout of the structure
U2-O2: To understand the process of excavation of the foundation
U2-O3: To realize the suitability of different tools for the different excavation operation
U2-O4: To determine the function and suitability of different types of foundation
U2-O5: To propose a suitable pumping method for dewatering

Unit-2 Outcome	EXPECTED MAPPING WITH COURSE OUTCOMES (1 - Weak Correlation; 2 - Medium Correlation; 3 - Strong Correlation)				
	CO-1	CO-2	CO-3	CO-4	CO-5
U2-O1	1	2	1	-	-
U2-O2	-	2	-	-	-
U2-O3		2	1	-	-
U2-O4	2	3	1	-	-
U2-O5	-	3	-	-	-

2.1 INTRODUCTION

In the previous unit, you got an idea about the various components of the building. We know that the foundation is constructed into two parts which are the substructure and superstructure. A substructure is a part that is below the ground. In this unit, we will learn about the construction of the substructure. The first step in the construction of the substructure is to plan the whole process of construction. The job layouts help in effective planning of the construction of the substructure.

2.2 JOB LAYOUT

Job layout can be defined as a drawing of the proposed construction site which shows different locations such as entry point, exit point, equipment, and material stores, temporary facility, site office, and place where workers will stay. Figure 2.1 shows the job layout of the building/mall under construction. Temporary facilities are not part of exact construction but they are made up to support specific tasks such as mixing plant, maintenance and fabrication shops, storerooms, etc.

Job layout helps us to plan the construction site in such a way that different construction resources like tools, machinery, materials, manpower, etc. can be arranged easily and optimal uses of construction space can be achieved. More complex the construction project or larger in scale detailed job layout is necessary as it helps in detailing and managing space on site. Following are the main purpose of job layout:

- It helps in reducing the completion time for the construction.
- It provides easy movement of equipment from one point to another.
- It helps in reducing wastage and deterioration of the material.

- It saves time by delivering and making a uniform flow of material at the site.
- It provides more safety to the worker at the site.
- The output from manpower and equipment can be increased using the job layout.

2.2.1 Factors Affecting Job Layout

A job layout is affected by numerous factors. The factors that affect a job layout are discussed as follows:

(1) **Type of project:** The type of project plays a very prominent role in making job layouts. The construction layout varies according to the project and its functioning. For example, if it is the construction of a mall then it requires a centrally located layout but if it is highway construction then it requires a number of the central layout at a suitable interval.

(2) **Method of construction:** The construction at the site may require a pre-cast structure or cast-in-situ or both. So, the job layout varies according to the requirements of the construction. For example, if there is a requirement for pre-cast structures, then a casting yard should also be present in the job layout otherwise there is no need of providing a casting yard.

(3) **Availability of resources:** Various resources such as manpower, machinery, material, etc. are required at the construction site. So, the job layout should have the provision for the different resources. The temporary arrangement of housing, food, or any other facility required should be given in the job layout. The location of the temporary facilities should be such that it remains safe from the adverse the impact of atmosphere.

(4) **Temporary road:** Roads are important to transfer material and equipment from one place to another place inside the construction site. It also provide access to the entry and the exit point from nearby road. The transportation facility should be given consideration in the job layout.

(5) **Miscellaneous facilities:** There may be other facilities that need to be shown in the job layout. The necessary provision for facilities such as electricity, water supply, material storage yard, and telephone connection should be there in the site job layout.

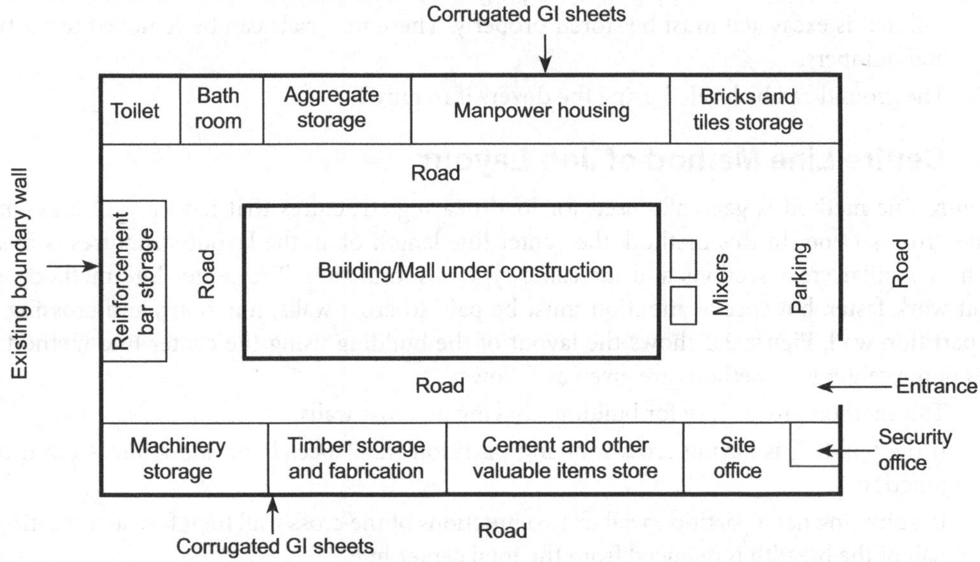

Fig. 2.1: Job layout of building/mall under construction

2.2.2 Site Clearance

Site clearance is the process taken to prepare a building site for its construction. It involves the clearance of vegetation, dead root mats, living or dead trees, and subsoil for a minimum depth of 30 cm from the ground surface. Site clearance also involves the work related to the demolition of any prior structure before the actual construction takes place. All depression occurs due to the removal of topsoil, vegetation should be filled with suitable material. It also involves the work related to leveling and making the ground ready for construction work to start.

Site clearance work is carried out very carefully and safely and all approval had to take before dumping the material taken out from the site. If any type of contaminated material such as asbestos or any hazardous materials is present on site, then an expert for the disposal of this type of material should be consulted before the disposal of hazardous material.

The following recommendations are considered during the process of the site clearance:

(1) The area to be cleared is marked properly.
(2) The trees and vegetation on the site are to be cleared off. The details of the trees to be cut should be listed and the trees and vegetation are removed manually or mechanically.
(3) The topsoil is excavated using the excavating equipment. The excavators are used for the removal of the soil.
(4) All the depressions and pits created during the process of clearing the trees and during the removal of the vegetation must be filled and compacted properly. The excavated soil is used for this purpose.
(5) The water can be used during the compaction of the pits. The density of the compacted pits should be near the surrounding area.
(6) If the structures exist in the area, these should be demolished mechanically. The demolition waste is dumped at the approved dumping sites, and the useful waste is stored properly. The soil that is excavated must be stored properly. These materials can be removed using tractors and dumpers.
(7) The ground can be leveled using the dozers if required.

2.2.3 Centre Line Method of Job Layout

The Centre line method is generally used for load-bearing structures that have a wall foundation of a similar cross-section. In this method, the center line length of all the layout structures is measured which have similar cross sections and the same type of foundation. The center line method helps in carryout work faster, but special attention must be paid to cross walls, junctions, and crossing points of the partition wall. Figure 2.2 shows the layout of the building using the center line method. Some features of the center line methods are given as follows:

- This method proves easy for buildings having no cross walls.
- If the building is having cross wall and partition wall, special consideration is given to their junction
- If a building has a partition wall or two junctions of the cross wall then for each meeting point half of the breadth is deduced from the total center line.

The following steps are followed during the layout using the center line method:

- First corner of the building is fixed by measuring the distance of one corner from the border of the plot and fixing it by using wooden pegs.
- From this wooden peg, the center line of the foundation is marked.
- Then half the width of the layout foundation is marked on either side of the center line by a wooden peg.
- Considering the orientation of the building, the center line of the foundation is matched with the baseline string.
- The perpendicular line is set using the first corner center line wooden peg
- Lime sand powder is used to mark the excavation line on either side of the center line.

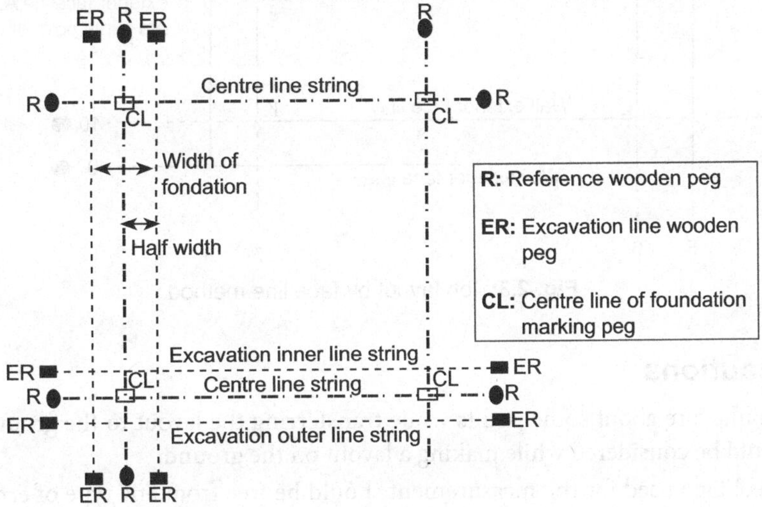

Fig. 2.2: Job layout by center line method

2.2.4 Face Line Method of Job Layout

In the face line method, the marking by lime sand powder is done on the wall exterior and interior surface. The marked line is extended by 2 meters away from the excavated point. Figure 2.3 shows the job layout by face line method. The different steps followed during the layout by the face line method are as follows:

- Step 1: Extended lines are to be marked for each wall exterior face by 1, 2, 3….so on.
- Step 2: The wooden pegs are used to mark this extended line for future reference as during excavation the face line will get disappeared.
- Step 3: The width of the wall between the wall face is excavated for the foundation.
- Step 4: Considering the orientation of the building, the face line of the foundation is matched with the baseline string.
- Step 5: The perpendicular line is set using the face line from the wall exterior marked by the wooden pegs
- Step 6: Diagonal check such as distance between AC, BD, EG, FH, and so on must be checked for the proper rectangular shape of the rooms.

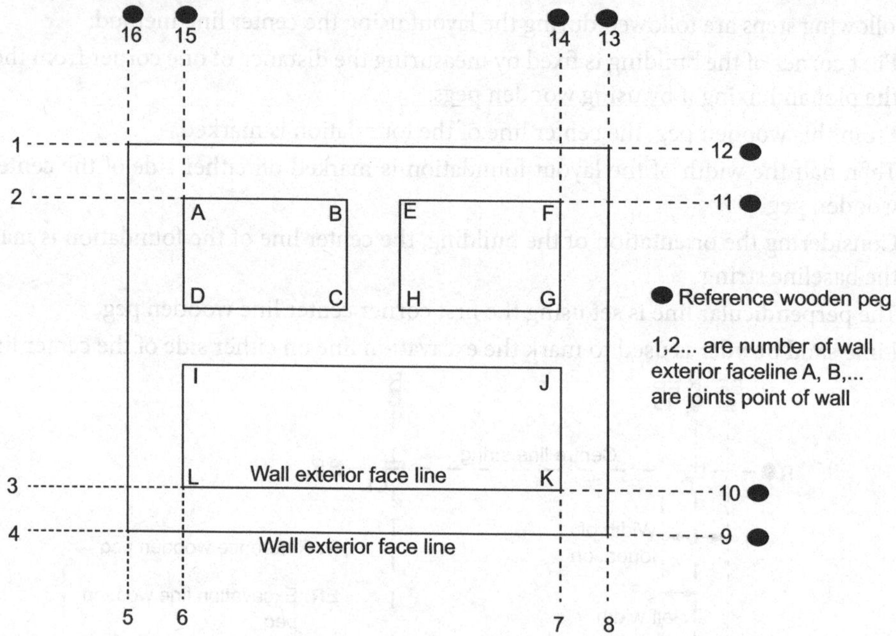

Fig. 2.3: Job layout by face line method

2.2.5 Precautions

The workers should care about some points while transferring the layout to the ground. The following precautions should be considered while making a layout on the ground:

- The steel tape used for the measurement should be free from any type of error. The steel tape should be checked for errors at regular intervals. As time passes, the steel tape can get elongated or shortened during the measurements. If there are errors in the tape, these should be corrected mathematically.
- The sketches of the location should be used with great care while fixing the boundaries of the plot.
- The dumpy level or the auto level should be used for the purpose of leveling. The level should be error-free and should be calibrated properly.
- The locations where the wooden pegs are to be installed must be properly marked. Any confusion can lead to errors in the job layout. The wooden pegs are kept on the same level and the markings on the wooden pegs should be clearly visible.
- The distance should be checked by at least two independent measurements.
- The rooms and the whole building should be checked for a diagonal check.
- The plumb bob should be straight while transferring the marking point on the ground. The inclination of the plum bob induces errors.
- The measurements of the distance, angles and elevations must be precise.
- The lime powder used for marking should visible clearly and distinct for proper excavation to avoid any confusion.

2.3 EARTHWORK

The process of the excavation and deposition of excavated material is known as earthwork. The earthwork includes the excavation, loading, transportation, deposition, compaction, and improvement of the soil. After the clearance of the site, the earthwork begins. Earthwork can be defined as engineering work that involves moving the soil or processing the soil from the earth's surface. In this process, the soil moves from one part to other parts and help in achieving the desired shape required for construction purpose. Earthworks mainly required heavy machinery equipment for the excavation of soil and backfill. Heavy machinery is generally used due to the involvement of large quantities of the material which has to be moved

2.3.1 Excavation for Foundation

Excavation can be defined as a process that helps in transferring subsoil, rock, or any other material using machinery and tools. It includes trenching, tunneling, earthworks, and wall shafts. One of the common practices using excavation is building construction. Excavation is primarily used in foundation for digging, trenching, and site development.

Excavation for small buildings is done manually using pick axes, spades, crowbars, etc. For deep foundations and for large buildings, excavation is carried out by using machinery for earth cutting. Setting out is done first to carry out excavation at the site. The Excavation is done on hard soil where the depth of the foundation is less than 1.5 m, trench sides don't require external support. In case of loose soil or deeper excavation, some type of support is required from preventing the sides to fall. Strutting and timbering are done continuously depending on the type of soil and foundation depth. In general, excavation trench width is kept equal to the width of foundation concrete, because wider excavation is not good for the strength and economy of the building as wider excavation is to be filled later by losing soil. The foundations should be checked for depth and width while excavating.

The depth of the foundation is checked using the boning rod. The boning rod height is adjusted to the depth of the foundation from the plinth level. The thread is tired to reference peg and the depth of the foundation is measured at different points by connecting the boning rod to the thread at various points. The width of the foundation is checked using timber plank equal to the width of the foundation and marking in the center.

The center line of the foundation is transferred from the timber plank to the bottom of the trench using the plumb bob. Now, the timber plank is placed at the center line to check excavation width by moving up and down inside the trench and the necessary width is provided where it is stuck. The following points are considered while doing excavation work:

- If the soil does not permit the vertical sides of the trench, support should be provided. The shoring can be done for the unstable sides of the trench.
- The excavated material should be placed at least 1 m away from the place of the excavation.
- The bottom of the trench should be perfectly leveled.
- The bed of the foundation should be rammed before pouring concrete for better stability.
- The soft soils or the rocks should be removed and the bed of the trench should be leveled and filled with stabilized soil.

- The wires should be placed around excavation work so that no person or any stray animal from the outside area falls inside the excavated region.
- Underground water pipelines, gas pipelines, communication cables, etc. should be protected while digging a trench.
- During excavation, if any archeological item is found, it is considered government's property and should be deposited at the government office.
- Special care is required while digging adjacent to old buildings so no damage is done to adjacent property
- Permission from the forest department is required to cut or transfer live trees from the excavation site, whose circumference is more than 30 cm and height is more than 1 m from the ground.
- The surplus soil left after filling the trench site should be disposed of properly.

2.3.2 Timbering and Strutting

Timbering and strutting are defined as temporary support to the trench when there is loose sub-soil or the depth of the trench is deep. It consists of timber plank and strut. According to Central Public Work Department (CPWD) manual, timbering is required for soft/loose soil if the depth of the trench is more than 2 m. In the case of hard and stable soil, the depth of the trench should be greater than 2.5 m. Timbering is also known as shoring. After the completion of foundation work, timbering is removed from the trench.

Some important terms used in timbering and strutting are as follows:

(1) **Polling Board:** Polling boards are generally wooden boards having a width of about 200 mm and thickness of about 45-50 mm. They are placed horizontal, vertical, or continuous along the length of the trench. The height of the board is kept equal to the depth of the trench. The polling board has direct contact with the soil and prevents it from falling into the trench. They are also known as sheeting.

(2) **Ranger or wale:** Rangers are placed behind the polling board in a horizontal position to support and keep the board in a straight standing position. Wales are the planks having a width of 200mm to 250 mm and a thickness of around 400 mm.

(3) **Strut:** Strut is timber plank that joins wale of either side to each other and kept wale intact in their position. Strut is sometimes directly attached to the polling board. Strut is placed horizontally across the trench. The dimension of the strut is 100 × 100 mm up to 2 m width trench and 200 × 200 mm for trench having 4 m width.

(4) **Bracing:** Bracing is provided diagonally to strengthen the timbering on either side of the trench.

The deep trenches require timbering and strutting generally. The process followed for timbering and strutting of the deep foundation is explained as follows:

(1) **Stay bracing:** This method is used when the soil is firm or hard to support the trench side. In this case, the depth of the foundation is not exceeding more than 2 m. In this method, the wooden planks are placed in the vertical position on both sides of the trench and connected with the help of two or more rows of the strut. An interval of 2 to 4 m is kept between the polling boards, extending to the depth of the trench. Figure 2.4 (a) shows the stay bracing for general soil.

(2) **Box Sheeting:** When there is loose soil present at the site of the trench, the box sheeting method is adopted. Box sheeting can only be used for depth of excavation up to 4 m. In this method, the vertical or longitudinal wooden planks are placed next to each other, and wales are provided in vertical or longitudinal rows to keep them in position. These wales are connected across with the help of struts. If the height of depth is more than 4 m bracings are provided along with struts. Figure 2.4 (b) shows the box sheeting.

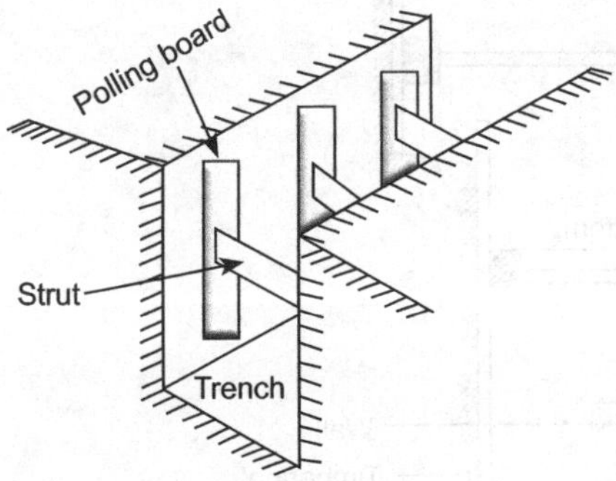

Fig. 2.4 (a): Stray bracings

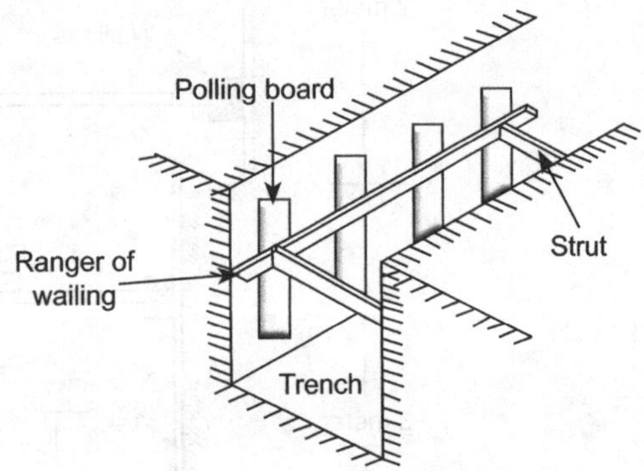

Fig. 2.4 (b): Box sheeting

(3) **Runner System:** This system is used when there is very loose and soft soil and required urgent support to the side of the trench as excavation progressed. This system is similar to box sheeting, except in this system long wooden planks are used in place of vertical sheets. Figure 2.4 (c) shows the runner system.

(4) **Vertical sheeting:** Vertical sheeting is done for trench having deep excavation up to 10 m in soft soil. In this method excavation is carried out in stages and timbering is done similar to the box sheeting. The width of the trench of the foundation decrease as the depth increases and offsets are provided at each stage.

Each stage is kept about 3 m the height and offset may vary from 250 to 500 mm per stage. Separate vertical sheeting is done for each stage supported by horizontal and vertical wale and struts. Figure 2.4 (d) shows the vertical sheeting.

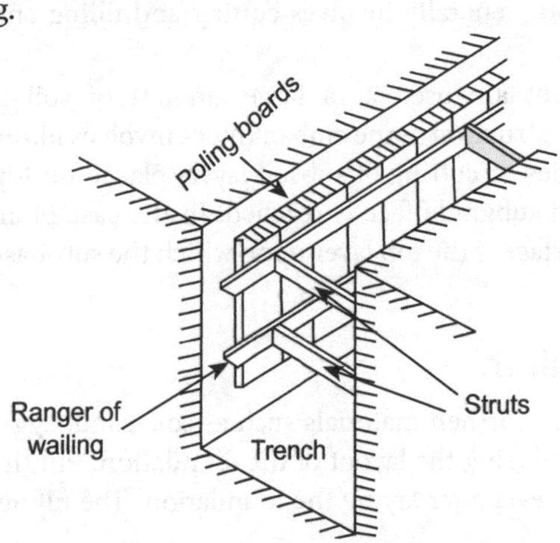

Fig 2.4. (c): Runner system

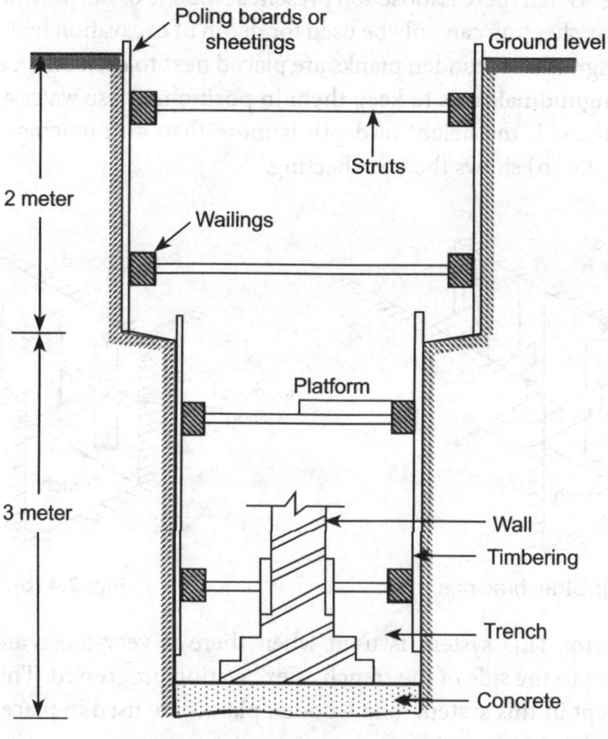

Fig. 2.4 (d): Vertical sheeting

2.3.3 Earthwork for Embankment

Earthwork is the engineering work using machinery and tools for moving or processing subsoil and underground rocks. The subsoil may be transported from one location to another or formed into any shape that is required for the construction. Earthwork generally involves cutting and filling and machine excavation at the site.

Heavy types of equipment are used as a large amount of soil is involved in the excavation. Construction of the embankment involves filling of subsoil in the form of a series or cutting of subsoil layers placed on top of each other until the desired subgrade face is reached. In the case of an embankment, the subgrade surface is the top layer upon which the sub-base will be placed.

2.3.4 Materials for Plinth

The filling is a process in which earthen materials such as soil, sand, aggregate, etc. are used to fill the depression or a hole produced during the layout of the foundation. Plinth filing is refers to the filing of excavated trench to the plinth level after laying the foundation. The filling material used for the plinth can be procure from outside or the material excavated from trench. The material should be free from dead trees, roots, lumps, organics materials or any type of harmful chemicals/salts. Filing material used for plinth can be procure from outside or the material excavated from trench.

Plinth that has to be filled should be cleared from all type of timbering and shuttering or any formwork. Space between foundation and side of trench filled up to the plinth level in layer. The thickness of layer should be kept 250 mm. Each layer should be watered and compacted by using rammers of weight 7-10 kg. The following materials can be used for the plinth filling:

(1) **Soil:** It should be free from any type of salts/chemicals and harmful organic matter. Black cotton soil is not used in filling trenches as this soil has expansion properties. Lump of earth soil of size more than 50 mm shall be broken or removed. Double handling of soil should be avoided and borrowed soil should be used directly to the fill excavated trench.

(2) **Murrum Soil:** Murrum is defined as fragments of rock but not rock. Murrum soil is red in color to the presence of iron oxide and is found in humid tropical or equatorial zones. It comes under laterite soil. The size of murrum soil used varies from finely powdered to 40 mm. its specific density is around 1.8 gm/cc.

(3) **Sand:** Sand used for filling the foundation up to the plinth level should be free from dust, organic matter, etc. Clay/silt quantity should be less than 5% in the sand filling.

2.3.5 Tools and Plants for Excavation

Earthwork equipment is typically heavy machinery and tools designed for excavation, digging the trench, and moving subsoil. Excavation of earth for the construction used hand tools and heavy equipment depending on the area of excavation and depth of the trench. Heavy machinery used for earthmoving driven by hydraulic mechanism. There can be hand tools or machinery tools.

Hand tools are driven by manpower and used for the excavation of small depths in small areas. Figure 2.5 shows some of the typical hand tools. The following hand-based tools are used for the excavation:

- **Spade:** It is a tool made up of a metal plate having sharp edges, connected with a wooden handle. Digging in the soil can be easily done with a spade as it has sharp edges.
- **Shovel:** It is used for digging and moving loose excavated soil. The shovel is similar to a spade, the only difference is its edge. It has higher curvature of the metal plate as compared to a spade and can move soil easily.
- **Hoe or Phaorah:** It is used for loosen and digging the soil and chopping weeds. It consists of a metal plate attached with wooden handle making acute angle at joint. It is widely used for small excavations.
- **Trowel:** It is a small size tool generally a size of hand used for small digging and removing top soil.
- **Pick Axe:** Pick axe is made up of hard spike which attached perpendicular to wooden handle. It is used for digging hard soil. The spike has wide blade on one side and pointed on other side.

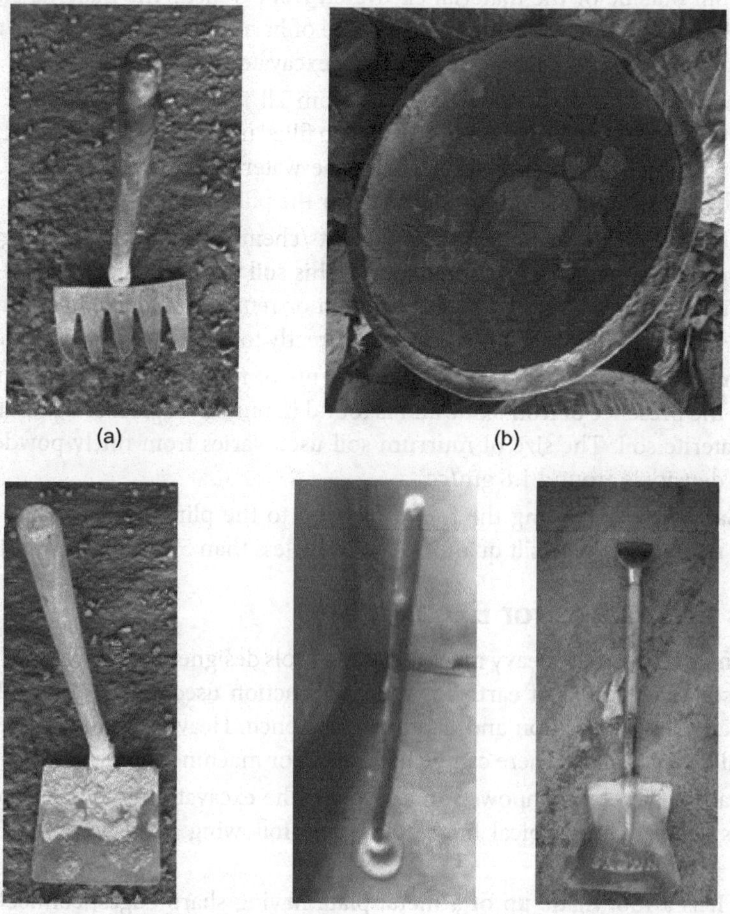

Fig. 2.5: Hand tools for excavation (a) Rake, (b) Basket, (c) Hoe or phaorah, (d) Rammer, (e) Shovel

- **Rammer:** It is made up of heavy metal piece attached with wooden handle having flat surface at the end. It is used for compacting the filling material in the trench.
- **Crow Bar:** It is made up of iron. It is used to plug the bricks from the ground.
- **Basket**: They are made up of plastics and galvanized iron. It is used to lift the loosen soil from small trench.
- **Rake**: it consists of horizontal wooden rod with metal teeth. It is used to clear the excavation site with tree debris and small rock.

Machinery tools are operated by mechanical and hydraulic force and are used to excavate larger depths. Figure 2.6 (a) and (b) shows machinery for excavation. Different types of machine tools used in modern-day construction are as follows:

- **Tracked Excavator:** A traced excavator is also known as a track hoe. It consists of a boom, dipper or stick, bucket, and cabinet on the rotating platform. The movement of the machine and function are done by using hydraulic cylinders and motors. It can rotate 360°. Traced excavators are moved by traction and can be used on uneven surfaces.

- **Backhoe Excavators:** It consists of a loader-style shovel/bucket at the front of the vehicle and a backhoe on the back side. By using this tool two operations can be performed that are digging and loading of the soil and lifting of the soil. It is smaller in size compared to other machine tools and is used for smaller construction activities. The wheel is used for moving the machine and can be moved to one place to another very easily.
- **Bulldozer:** It is made up of a heavy vehicle having sharp edges of a hard steel plate at its front. It is used for digging and lifting the excavated soil by using a sharp edge. Hydraulic cylinders and motors are used to raise and lower the metal plate. A bulldozer is also known as a dozer and is available in a wheel and tracked form. The readers can scan the code to see more picture of excavation machines.

Fig. 2.6 (a): Tracked excavator

Fig. 2.6 (b): Backhoe

2.4 FOUNDATION

The word 'foundation' has emerged from the Latin word 'fundare'. The word 'fundare' means to 'set on a solid base'. Foundation is the lowermost part of the structure which transfers the load of the superstructure and other loads to the ground. A foundation may be defined as a solid base on which the structure rests. Foundation remains in direct contact with the ground to transfer the dead load and superimposed load. The foundation plays a major role in the stability and strength of the structure. As the ultimate load from all the building components is to be transferred through the foundation, it is essential to construct the foundation carefully. The selection of type of foundation requires deep knowledge about the soils and their characteristics Fun

2.5 FUNCTIONS OF FOUNDATION

The foundation is one of the most important parts of the building. The structural elements like columns, beams, and slabs transfer the load to each other but it is the foundation that ultimately transfers the load of the whole structure to the earth. So, the foundation plays a vital role in the safety and serviceability of a structure. The major functions of the foundation are discussed as follows:

(1) A foundation provides a hard and level surface that is suitable for the construction of a superstructure over it.

(2) The foundation distributes the load from the structure to a larger area which helps in reducing the load intensity. We know that pressure is the ratio of force to area. If the area is larger the pressure reduces. The foundation acts on the same principle. The load of the structure

is distributed by the foundation and transferred to the ground. The larger the area of the foundation, the lesser will be the load intensity and the safer will be the structure.

(3) The load is distributed uniformly by the foundation. If the load is distributed non-uniformly or unevenly, there can be differential settlements in the base. The differential settlements can cause cracks in the building components. In extreme cases, even failure of building components can occur.

(4) There is a significant impact of winds and earthquakes on the buildings. The wind and earthquake loads have a tendency to overturn the building due to their horizontal impact. The foundation binds the superstructure and provides lateral stability to the structure.

(5) The foundations save the components of the building from the cracks during the movement of the soil. The soils can expand or contract which can cause the cracking of the structural components of the buildings. Special measures during the construction of the foundation are taken to save the structure from cracking.

(6) The foundation reduces the impact of undermining and scouring on the building.

2.6 TYPES OF FOUNDATION

The foundations are classified based on their dimensions. The foundations can be classified into two types which are the shallow foundation and deep foundation. A shallow foundation may be defined as a foundation that has a depth equal to or lesser than its width. If the depth of the foundation is more than its width, then it is known as a 'deep foundation'. The shallow and deep foundations can be classified further. The shallow and deep foundations and their types are explained in the subsequent sections.

2.7 SHALLOW FOUNDATION

A footing is the part of the foundation that transfers the load to a larger area. The base of the column or wall is enlarged to increase the area at the bottom, known as footing. Foundation is the combination of footings. Shallow foundations can be further classified into the following four categories:

- Spread footings
- Combined footings
- Strap footings
- Mat foundations

2.7.1 Depth of Shallow foundation

The soil at the surface is loose and weak generally. So, the topsoil is removed for the construction of the foundation. There are the following three choices are considered:

- If the topsoil is very weak in nature and has poor strength, it will not be able to withstand the load from the foundation. In this case, the soil can be treated and its mechanical properties can be enhanced by the process of stabilization of soil. The loose soil of a larger area can be replaced with a mixture of gravel and sand.
- The topsoil can be removed and replaced with plain concrete. It will enhance the strength of the base.
- If the soil at the surface has enough strength to withstand the loads from the foundation, still a minimum depth of the foundation should be kept. The minimum depth of a shallow foundation can be calculated from Rankine's formula as follows:

$$Dmin = \frac{P}{\gamma}\left(\frac{1-Sin\varphi}{1+Sin\varphi}\right) \tag{2.1}$$

Here P is the pressure at the footing's base in kN/m²
γ is the unit weight of the soil in kN/m³
φ is the angle of repose

The angle of repose represents the inclination at which a particular material can stand safely. For example, dry sand has an angle of repose from 25° to 35°.

2.7.2 Spread footings

As the name suggests, the spread footings are the footings that distribute the load to the earth by spreading the load over a larger area. The area at the base of columns or walls can be spread in different ways. Based on the different shapes of the spread footings, the spread footings can be further divided into the following six types:

a. Single column footing
b. Stepped column footing
c. Sloped column footing
d. Strip wall footing
e. Stepped wall footing
f. Grillage footing

a. Single column footing

A single footing is given at the base of the column. The area of the column is spread by a single concrete base. Figure 2.7 (a) and (b) show the single footing. During the construction of the single-column footing, the following recommendations can be considered:

- If the load coming to the column is 'P' and the safe bearing capacity of the soil is 'qs', the area of footing is given by the following formula:

$$Area\ A = \frac{P}{qs} \tag{2.2}$$

- The shape of the concrete pad can be same as the shape of the column. In figures 2.7 (c), x shows the offsets. The offset of 10 cm to 20 cm can be given. The dimensions of the base pad can't be less than two times of the column dimension. Figure 2.7 (d) shows details of plan.

Fig. 2.7 (a): Single column footing

Fig. 2.7 (b): Plan of single footing

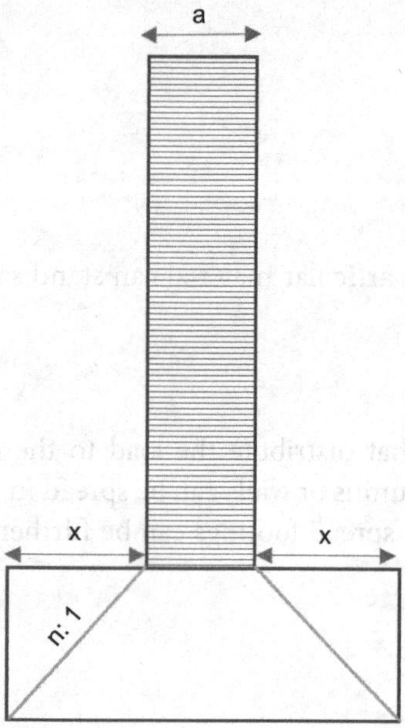

Fig. 2.7 (c): Single column footing

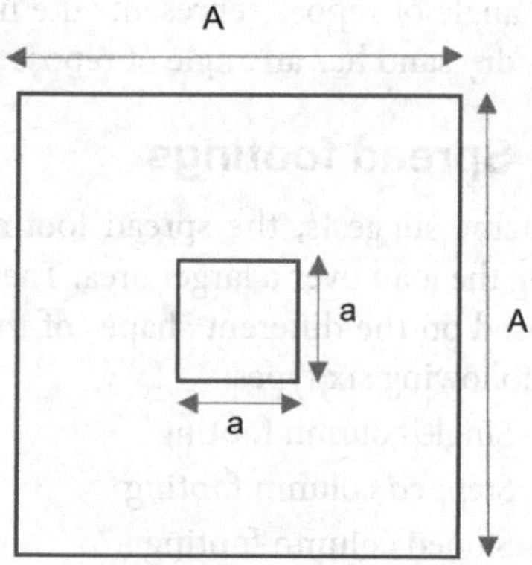

Fig. 2.7 (d): Details of plan of single footing

- ➢ If the pad is constructed of cement concrete, thickness of the pad should be equal to the offset. If the pad is constructed from lime concrete, thickness of the pad should be 1.5 times of the offset.
- ➢ The spread is shown by n:1. The angle of spread should be 1:1 for the cement concrete footing while angle of spread for the lime concrete should be 0.66:1.
- ➢ The single column footings provided only of the load coming to the column is small and the bearing capacity is large.

b. Stepped column footing

If the load to be distributed in very high, the stepped footing can be provided for the columns. In the case of stepped footing, the area at the base of the column is increased by providing the steps with a gradual increase in their area. Steps are shown in different colours for better understanding. Figure 2.8 (a) and figure 2.8 (b) show the stepped footing for the column.

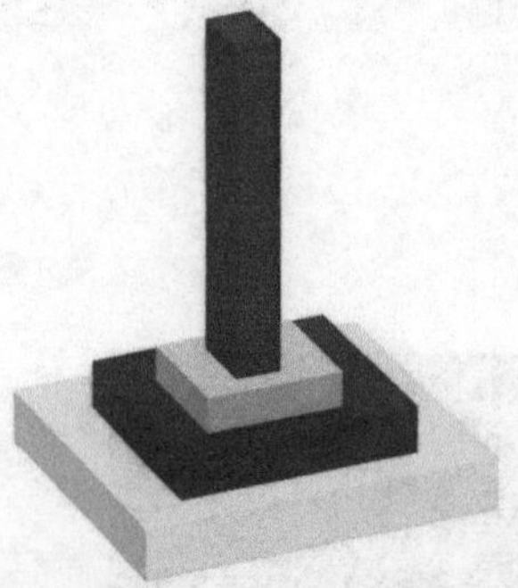

Fig. 2.8 (a): Stepped column footing

Fig. 2.8 (b): Plan of stepped column footing

During the construction of stepped column footing, the following recommendations are considered:

- The stepped column footing is provided when the load coming to the column is more than the safe bearing capacity of the soil.
- The spread is given so that the load can be transferred gradually to the base.
- Consider figure 2.8 (c), in the stepped footing, the layers of masonry are provided before giving the concrete base. The spread is shown with n:1 and n1:1. The rate of spread for the masonry is kept 0.5:1, 0.66:1 for the lime concrete, and 1:1 for cement concrete is provided.
- For the masonry pillar, the offset should be between 10 to 15 cm and the step height may vary from 15 cm to 22.5 cm.

c. **Sloped column footing**

The sloped column footing is constructed on an inclination of 45° from all sides. Sloped footings save material and gradually transfers the load to the ground. The sides of the footing are provided with a slope with a concrete layer at the bottom for uniform distribution of loads.

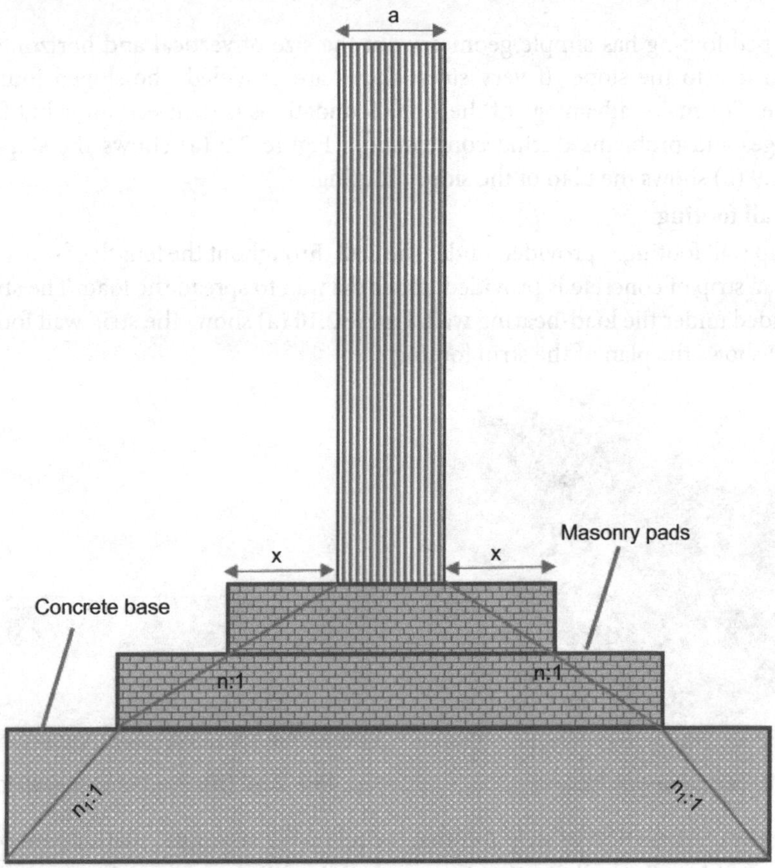

Fig. 2.8 (c): Elevation of stepped column footing

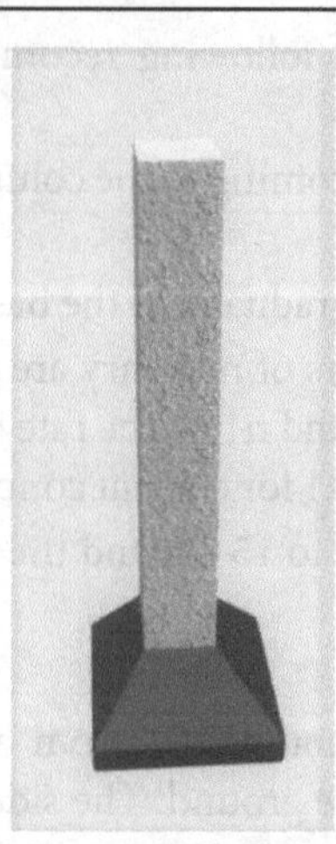

Fig. 2.9 (a): Sloped column footing

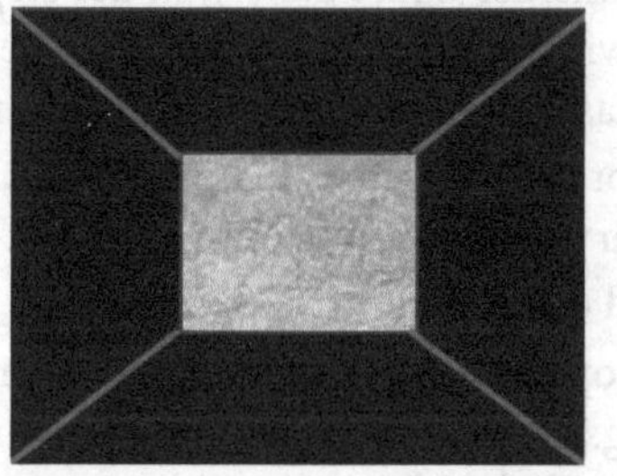

Fig. 2.9 (b): Plan of sloped column footing

The sloped footing has simple geometry but the size of vertical and horizontal bars can't be uniform due to the slope. If very steep slopes are provided, the sloped foundation can be unstable. The major advantage of the slope foundations is their economy but this poses more challenges and problems during construction. Figure 2.9 (a) shows the sloped footing and figure 2.9 (b) shows the plan of the sloped footing.

d. **Strip wall footing**

The strip wall footing is provided under the wall throughout the length of the wall. In strip wall footing, a strip of concrete is provided under the wall to spread the load. The strip wall footing is provided under the load-bearing wall. Figure 2.10 (a) shows the strip wall footing and figure 2.10 (b) shows the plan of the strip footing.

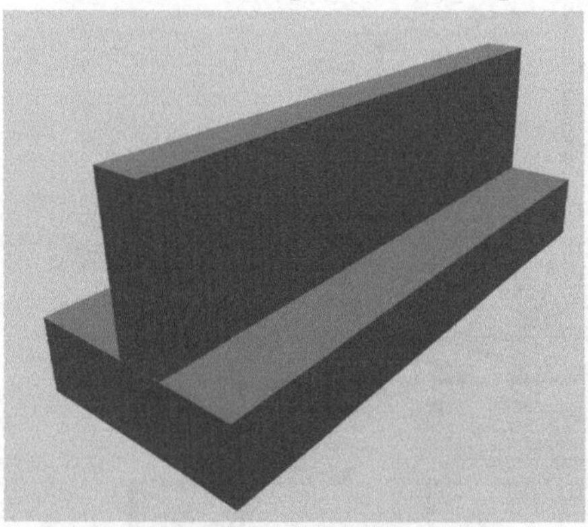

Fig. 2.10 (a): Strip wall footing

Fig. 2.10 (b): Plan of strip wall footing

During the construction of strip footing, the following recommendations are considered:

- The width of the wall footings is calculated the following formula:

$$B=\frac{P}{qs} \tag{2.3}$$

Here, P is the load coming to the footing and qs is the safe bearing capacity of the soil

- When the safe bearing capacity of the soil is very high and the load coming to the wall is light, in this case, simple strip footing is provided.
- The concrete pad is provided directly under the wall and there is no need of providing the masonry offsets.
- The offsets along the wall should be from 10 cm to 20 cm.
- The angle of spread is kept the same as in the case of simple column footing.
- The width of the concrete pad is kept equal to or more than two times of the wall width.
- The depth of the concrete pad should be equal to the offset in the case of cement concrete and should be equal to 1.5 times of offset in the case of lime concrete.

e. **Stepped wall footing**

The area of the base of the wall is gradually increased in the form of steps. If the loads are very high on the load-bearing walls, stepped wall footing can be constructed. Figure 2.11 (a) shows the stepped wall footing and figure 2.11 (b) shows the plan of the stepped footing.

Fig. 2.11 (a): Stepped wall footing

Fig. 2.11 (b): Plan of stepped wall footing

During the construction of stepped wall footing, the following recommendations are considered:

- If the load on the wall is very high and the bearing capacity of the soil is not significant, in this case, stepped wall footing can be used. When the load is very high on the wall, the masonry offsets are provided as in the case of the column.
- The maximum angle of the spread in the wall step footing can be given 0.5 in horizontal to 1 in vertical in the case of brick or stone masonry.
- The maximum angle of spread should not be more than 0.66 in horizontal to 1 in vertical in the case of lime concrete and it should be 1:1 for cement concrete.
- The minimum depth is calculated using the following formula:

$$D = \frac{1}{2n}(B - T) \tag{2.4}$$

Here, n is the spread in horizontal with respect to vertical
B is the width of the footing
T is the width of the wall

- In the case of brick masonry, the offset should be equal to or less than 5 cm and the height of the step can't be more than 10 cm.

- For stone masonry, the offset may vary from 7.5 cm to 10 cm while the height of the step may be from 15 cm to 20 cm.
- The depth of the concrete bed can be calculated using the following formula:

$$Dcmin = a\sqrt{\frac{3q}{m}} \tag{2.5}$$

Here, a is the offset of concrete beyond masonry in figure 2.11 (c)

m is the modulus of rupture for concrete

q is the net soil bearing capacity

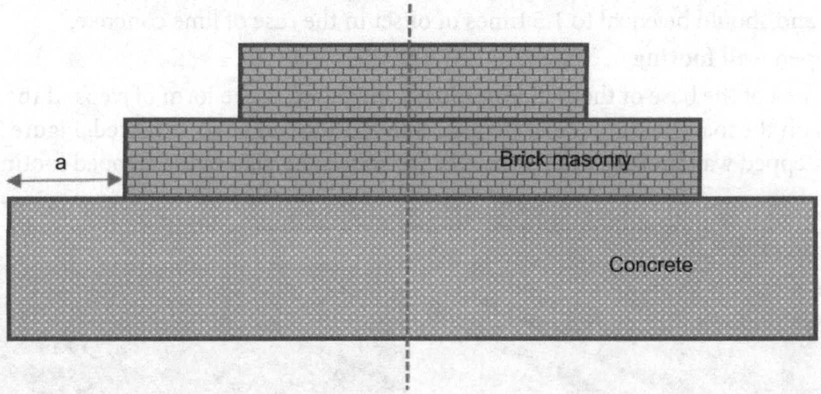

Fig. 2.11 (c): Elevation of steps in stepped footing

f. **Grillage foundation**

A grillage foundation consists of a combination of beams laid over each other resting on the layer of concrete. The area is spread at the bottom of the column by the combination of beams to spread the superimposed loads. The grillage foundation provides more stability and strength and can take very high loads. The grillage foundation can be constructed from steel or timber. Figure 2.12 (a) shows the grillage foundation and figure 2.12 (b) shows the plan of the grillage foundation. A grillage foundation can be constructed from steel or timber. If the superstructure consists of very heavy elements like steel columns, piers, etc., the grillage foundation can be recommended. The major advantage of the grillage foundation is that it can be constructed in lesser time as prefabricated parts are to be installed. It can take and distribute heavy loads efficiently to the ground.

During the construction of the grillage footing, the following recommendations are considered:

- The grillage foundation is provided for the heavy load of columns or steel stanchions if the bearing capacity of the soil is poor.
- The depth of the grillage foundation is kept from 1 m to 1.5 m.
- In the steel grillage foundation, the steel beams are separated by pipe separators of 25 mm diameter and spacers of 20 mm diameter. The distance between the steel beams should be sufficient so that concrete can get compacted easily. The minimum clearance between the beams can be kept 8 cm.
- The minimum concrete cover of 15 cm is provided to the lower beams and 10 cm for the upper beams.

➤ In case of timber grillage foundation, the timber boards of 8 m to 10 cm thickness are stacked to form the timber grillage. The maximum spacing between the timber beams can be 38 cm.

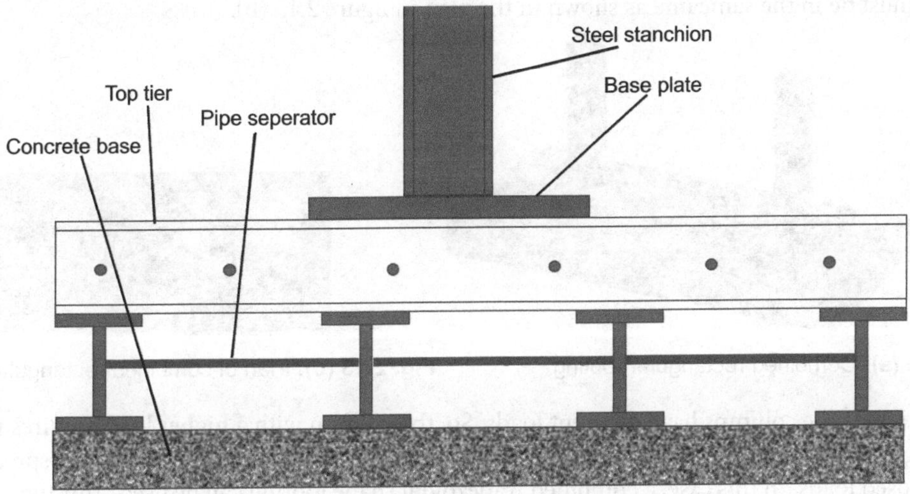

Fig. 2.12 (a): Grillage foundation

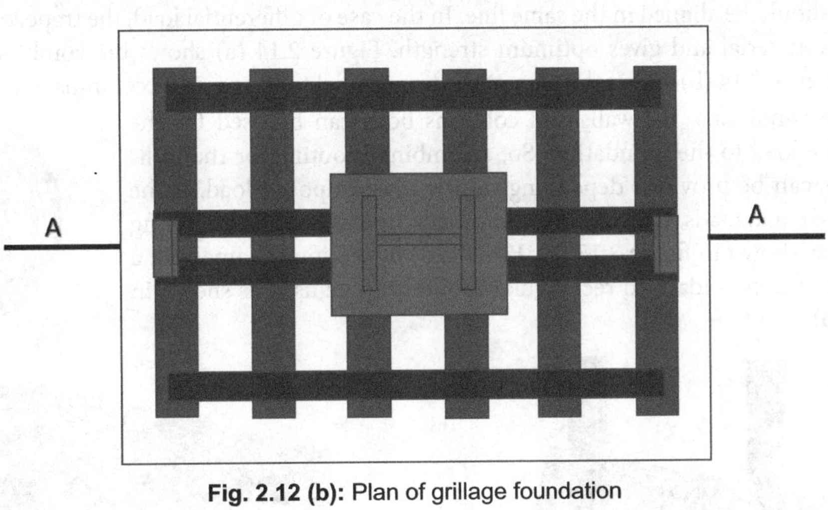

Fig. 2.12 (b): Plan of grillage foundation

2.7.3 Combined footings

Sometimes if the columns are constructed near to each other, it is not possible to provide the individual footings. So, a combined footing for two or more columns is provided. The combined footings are more economical as compared to the individual footing for each column. The combined footing can be of the following three types:

- Rectangular combined footing
- Trapezoidal combined footing
- Combined column wall footing

If the load from the columns is equal to each other, a rectangular column footing is provided. The rectangular column footing takes the load from the column and spread it evenly on the ground. Figure 2.13 (a) shows the rectangular combined footing. The centroid of footing and the centre of gravity of the columns must be in the same line as shown in the plan in figure 2.13 (b).

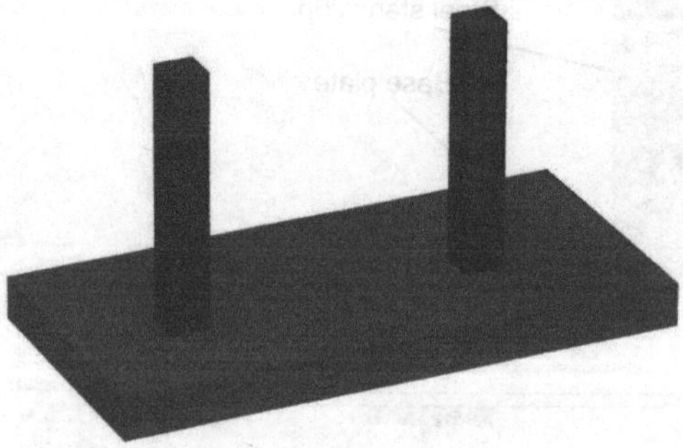

Fig. 2.13 (a): Combined rectangular footing

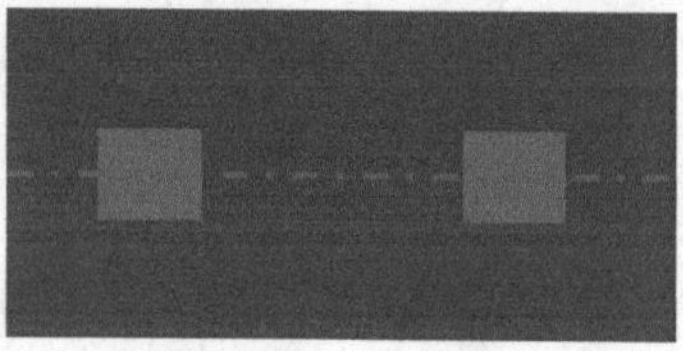

Fig. 2.13 (b): Plan of combined rectangular footing

Sometimes the columns have different loads. So, the column with a higher load requires more area to spread the load while the column with lower loading requires a lesser area for spreading the superimposed loads. In this case, a combined trapezoidal shape footing can be used. This type of footing is known as combined trapezoidal footing. The centre of gravity of the columns and the centroid of the footing area should be aligned in the same line. In the case of differential load, the trapezoidal combined footing saves material and gives optimum strength. Figure 2.14 (a) shows the combined trapezoidal footing and figure 2.14 (b) shows the plan of the trapezoidal footing for the columns.

In some conditions, the walls and columns both can be used for the transfer of the load to the foundation. So, a combined footing for the walls and columns can be provided depending on the superimposed load. If the columns have equal loads, the combination of rectangular combined footing can be used as shown in figure 2.15 (a). If the load by columns is unequal, a combination of trapezoidal and rectangular footing can be used as shown in figure 2.15 (b).

SCAN ME

Fig. 2.14 (a): Combined trapezoidal footing

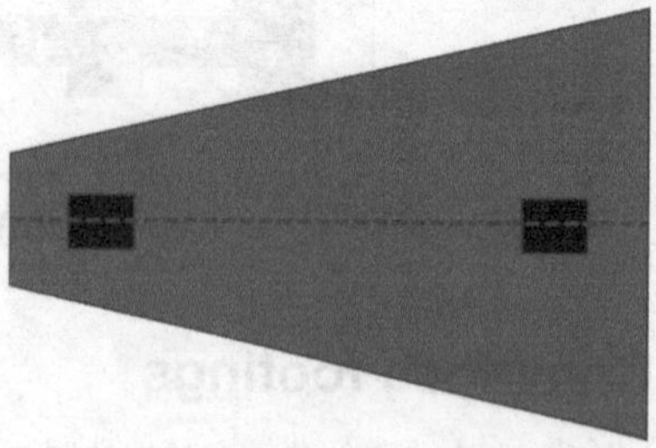

Fig. 2.14 (b): Plan of combined trapezoidal footing

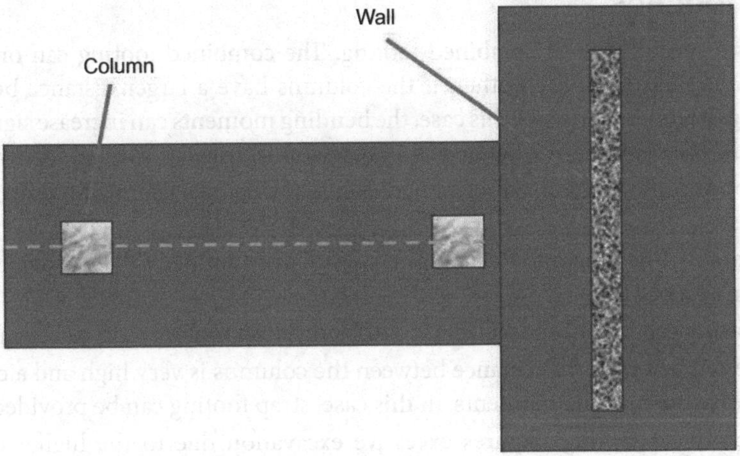

Fig. 2.15 (a): Rectangular column wall footing

The following recommendations are considered in the case of the combined footings:

- The combined footing acts as a beam that is loaded in an upward direction and rests on columns. The portion of the beam outside the columns acts as a cantilever.
- The load of the column and weight of the footing are added to get the total load in the downward direction.
- The area of the footing is calculated by dividing the total downward load by the safe bearing capacity of the soil.
- The location of the centre of the columns from the edges is calculated such that the centre of gravity of the loads coincides with the centre of gravity of the footing.
- The shear force and bending moment acting on the combined footing is calculated and checked while designing.
- The reinforcement is provided to counteract the induced bending moment and checked for safety.

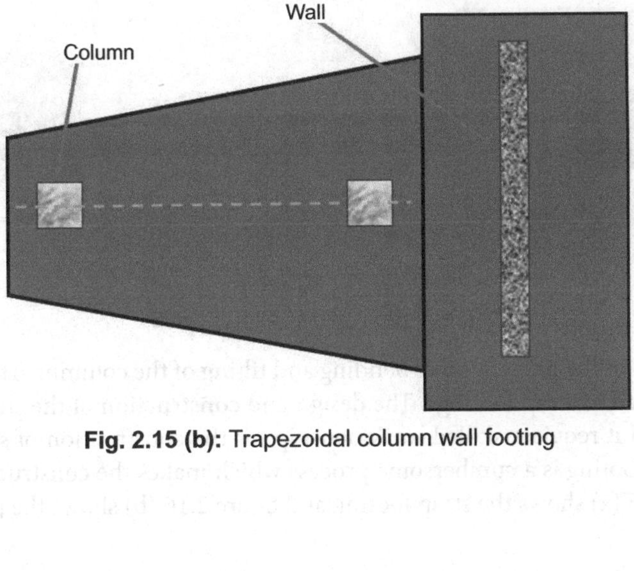

Fig. 2.15 (b): Trapezoidal column wall footing

2.7.4 Strap footings

A strap footing is a special type of combined footing. The combined footing can only be used if the columns are not much far from each other. If the columns have a larger distance between them, the trapezoidal footings become narrow. In this case, the bending moments can increase significantly, leading the structure to fail. The strap may be defined as a beam connecting the column's individual footings.

So, in strap footing, the individual footings are combined using a beam. The strap footing is used in the following cases:

a. If the property line is near to the column and isolated footing can't be provided, a strap footing can be provided.

b. If the bearing capacity of the soil is not sufficient, strap footings can be recommended.

c. As suggested before, if the distance between the columns is very high and a combined footing can increase the bending moments, in this case, strap footing can be provided.

d. If the combined footing requires excessive excavation due to the higher distance between columns, it can prove costly. So, strap footing can be a cost-effective solution in this case.

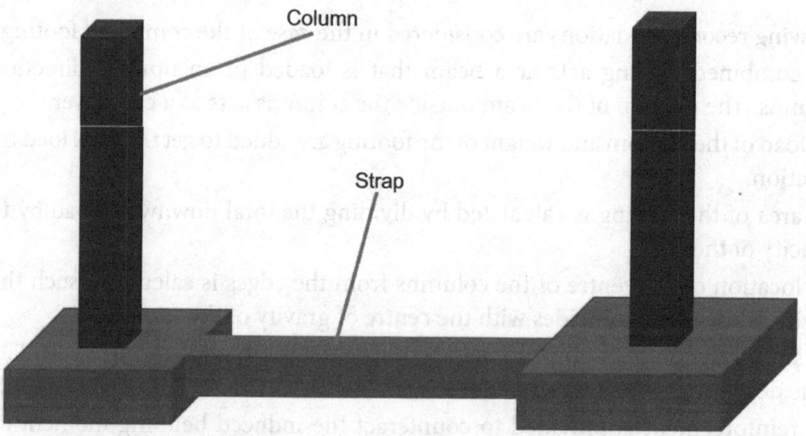

Fig. 2.16 (a): Strap footing

Fig. 2.16 (b): Plan of strap footing

Strap footings help in the prevention of bending and tilting of the columns. The loads and moments are uniformly distributed in strap footings. The design and construction of the strap footings is a time-consuming process and it requires skilled workmanship for the construction of such foundations. The designing of the strap footing is a cumbersome process which makes the construction of strap footings difficult too. Figure 2.16 (a) shows the strap footing and figure 2.16 (b) shows the plan of strap footing.

2.7.5 Raft foundations

The raft foundation is also known as the mat foundation. The raft foundation is provided throughout the base of the superstructure as a mat. It is a continuous slab that supports a number of columns. The mat foundation transfers the load of all the components of the building to the earth. The raft foundation also reduces the differential settlements.

The raft foundation is suitable for the regions where the bearing capacity of the soil is low or the soil is subject to movement. In such cases, a raft foundation is one of the most suitable types of shallow foundation. Sometimes structures like commercial buildings, water tanks, silos, storage tanks, etc. have very high loading. So, a mat foundation can be used when the structure faces very high superimposed loads. Figure 2.17 (a) and (b) shows the raft foundation.

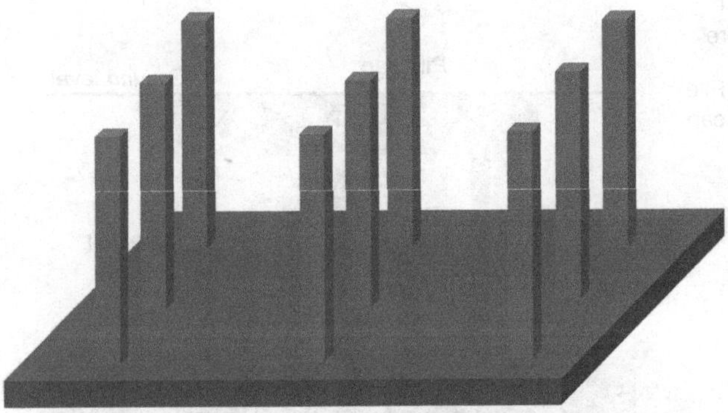

Fig. 2.17 (a): Raft footing

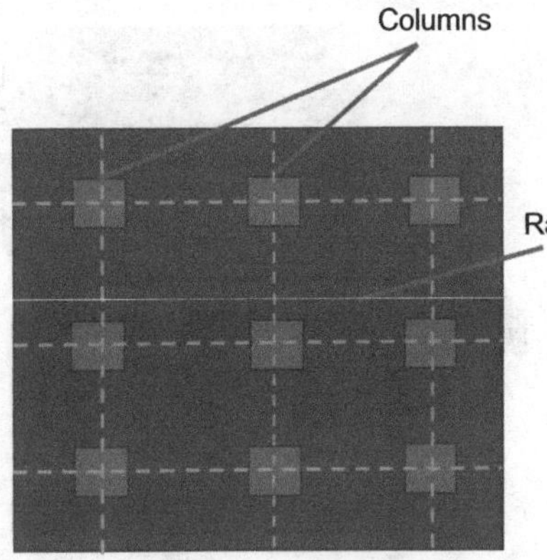

Fig. 2.17 (b): Plan of raft footing

The following recommendations are considered in the case of a raft footing:

- A raft footing covers the whole area under the superstructure and the components of the building rest on the raft.
- The raft is designed as a reinforced flat slab.
- The raft foundation reduces the differential settlements. If the soils are highly compressible and have chances of high differential settlements, the raft foundation can be provided.
- The self-weight of the raft footing is neglected in the design.
- The raft should be provided with a suitable waterproof membrane and cured properly after laying the concrete.

2.8 DEEP FOUNDATION

We know that the shallow foundations transfer the load to the hard stratum below the ground by means of footings. Sometimes the hard stratum under the ground lies at a significant depth. In this case, it is not feasible to provide a shallow foundation. If the hard stratum lies deep below the ground, a deep foundation can be provided. A deep foundation is the type of foundation that has more depth than its width. Deep foundations are of the following three types:

- Pile foundation

- Pier foundation
- Well foundation or caissons

2.8.1 Pile foundation

A pile may be defined as a slender long column made of timber, concrete, steel, or composite material that is used to transfer the load of the structure through its bottom or friction action or by a combination of both. The diameter of the piles is generally equal to or less than 0.6 m.

Based on the mechanism of transfer of the load, the piles may be classified further into the following three types:

(1) End bearing piles
(2) Friction piles
(3) Combined piles

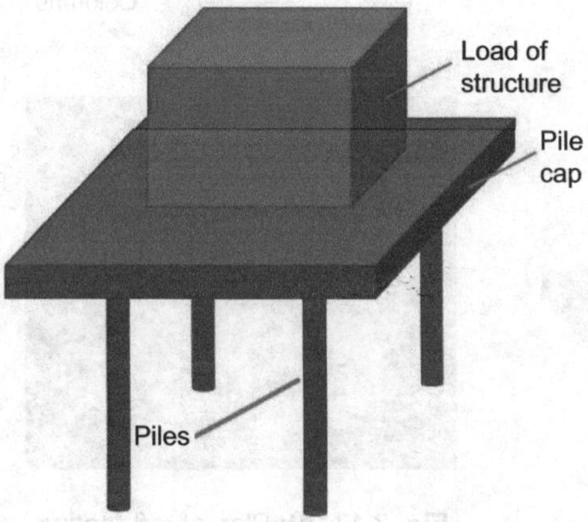

Fig. 2.18 (a): Pile foundation

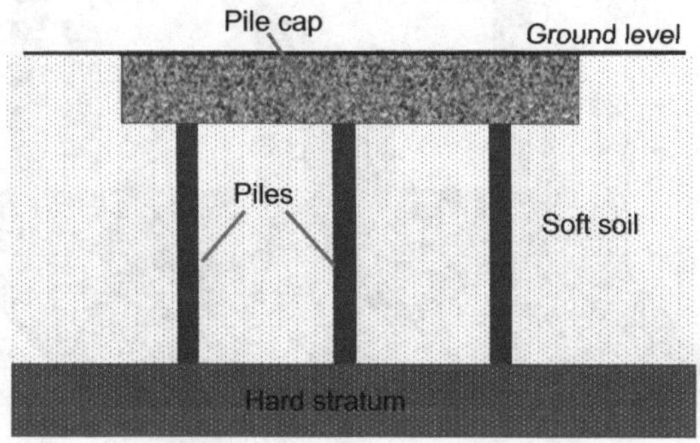

Fig. 2.18 (b): Mechanism of end bearing piles

The end-bearing piles transfer the superimposed load through their bottom tip. These piles act as a load-bearing structure. These types of piles are used to transfer the load through the layer of water or soft soil to an underlying hard stratum. Figure 2.18 (a) shows the pile foundation and figure 2.18 (b) show the mechanism of transfer of load for the end-bearing piles.

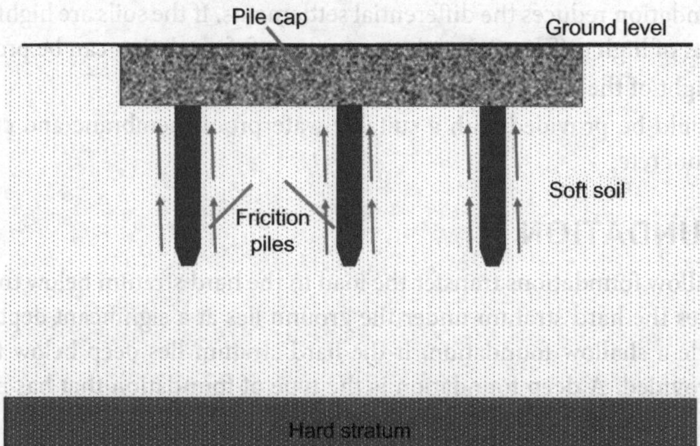

Fig. 2.19: Friction piles

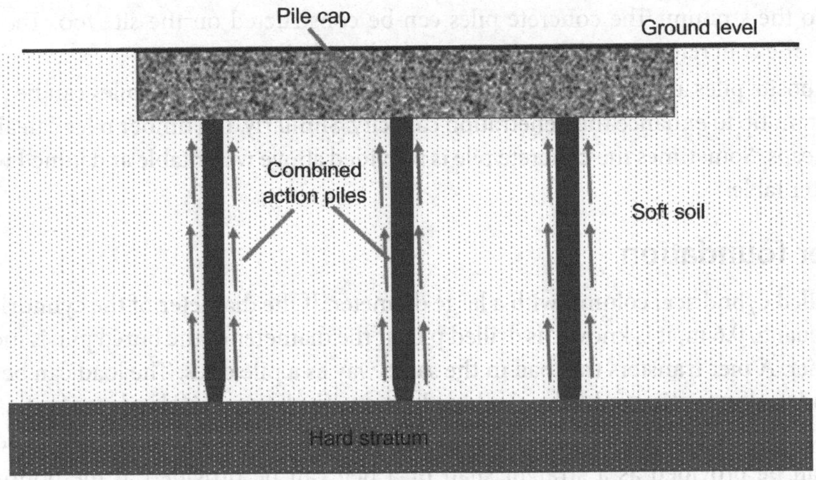

Fig. 2.20: Combined action piles

Sometimes the hard stratum may lie under extreme depth and taking the piles to the required depth may prove uneconomical. In this case, friction piles can be used. The friction piles transfer the superimposed load by their skin friction. The perimeter of the pile remains in contact with the soil and transfers the load to the soil around the circumference of the piles. Figure 2.19 shows the transfer of the load by the skin friction mechanism.

The combined piles transfer the load by the combined action of their bottom tip and body friction. These types of piles have greater bearing capacity. Normally combined action piles are used in the case of granular soil. Figure 2.20 shows the combined action piles.

The piles can be made of different materials. The piles can be classified into the following four categories based on the material:

(1) Timber piles
(2) Steel Piles
(3) Concrete Piles
(4) Composite piles

The trunks of the trees are used for making the timber piles. The timber used for making these piles should be straight and has sufficient strength. The timber should be defect free. The major advantage of these piles is that the timber is easily available. The cost of this type of pile is less. The timber piles have lesser life as these are prone to termite attack. So, these types of piles are used for underwater construction. The bottom of the timber piles is provided with a steel cap so that it can be protected from damage.

The steel piles are strong in tension. These are provided in the form of pipes or H section. The steel piles are durable. These can be driven into the ground. These types of piles are easy to transport and can withstand rough handling. The bearing capacity of these piles is very high. If the ground conditions are very stiff and penetration is difficult, the steel piles can be a good alternative. These piles can be prone to erosion. Epoxy coating is provided for saving the steel piles from erosion.

The concrete piles have sufficiently strong and can withstand very heavy loads. The cement concrete is used to construct the concrete piles. These types of piles are either precast type or cast in situ type. The precast type concrete piles are prefabricated in plants and transported to the site. These piles can be driven into the stratum. The concrete piles can be constructed on the site too. The bore holes are excavated and the concrete is poured to construct the bored piles.

The composite piles are made of more than one material. The concrete piles can be provided a steel cap at the bottom tip to enhance the penetration capacity. Similarly, the timber piles can also be provided with the steel caps. Sometimes the concrete pile is provided above water table and joined with timber pile below the water table.

2.8.2 Pier foundation

A pier is a drilled cylindrical column with a large diameter. If the diameter of the cylindrical structure is less than or equal to 0.6 m, it is known as drilled pile. If the diameter of the bored pile is more than 0.6 m, it is called a pier. A pier transfers the load in the same way as a pile does. The load can be transferred by the bottom end or skin friction or by the combined action. The heavy axial loads are vertically transferred by piers. A pier may be used instead of a number of piles which leads to the ultimate economy.

A pier can be provided as a straight shaft or a bell can be provided at the bottom. If the hard stratum lies within 5 m below the ground level, a straight concrete pier can be used for the transfer of the superimposed load. If the hard stratum is deeper than 5 m, the bottom of the pier can be enlarged in the form of a bell. The angle of the bell portion is kept around 60°.

The drilled piers require strict supervision during the concreting. The construction of the piers requires highly skilled workmanship. There is a need of conducting extensive subsurface explorations before the construction of piers. The excavation is to be done very carefully during the construction of piers as there is a chance of subsidence or soil failures. Figure 2.21 (a) shows the straight shaft pier and figure 2.21 (b) shows the bell-shaped pier.

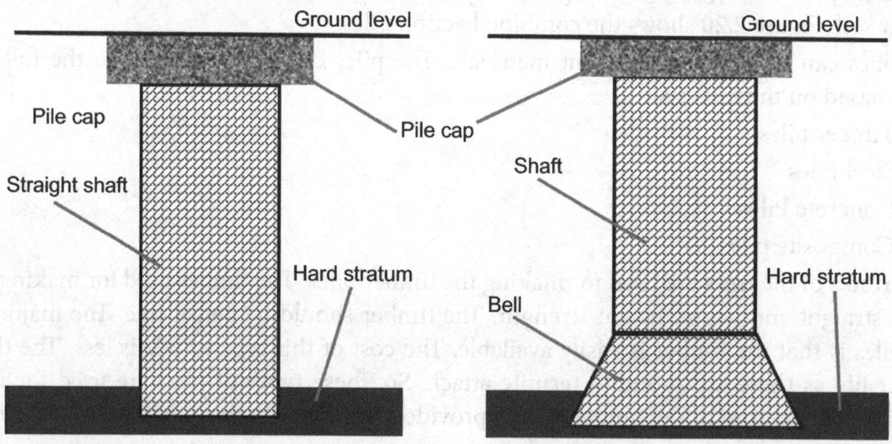

Fig. 2.21 (a): Straight shaft pier **Fig. 2.21 (b):** Bell-shaped pier

2.8.3 Caissons

Caissons are box or cylindrical type structures that are hollow in nature. The caissons are fabricated on the ground and penetrated at the waterbed. The caissons are used to support the bridges in the water bodies. The caissons are of the following three types:

- Open caissons or well foundation

- Floating caissons
- Pneumatic caissons

The open caissons are also known as well foundations. The open caissons remain open from the top and the bottom. The bottom of the open caissons has penetrating cutting edges which help it in penetrating through the bottom of the river, canals, harbours, or other water bodies. During underwater construction, the biggest challenge is to keep the water away from the construction site. Open caissons can fulfill this purpose too. The open caissons help in the construction of piers under the bridges. The open caissons are used for the construction on soft soils. These can be made of steel or concrete. Figure 2.22 shows the open caisson.

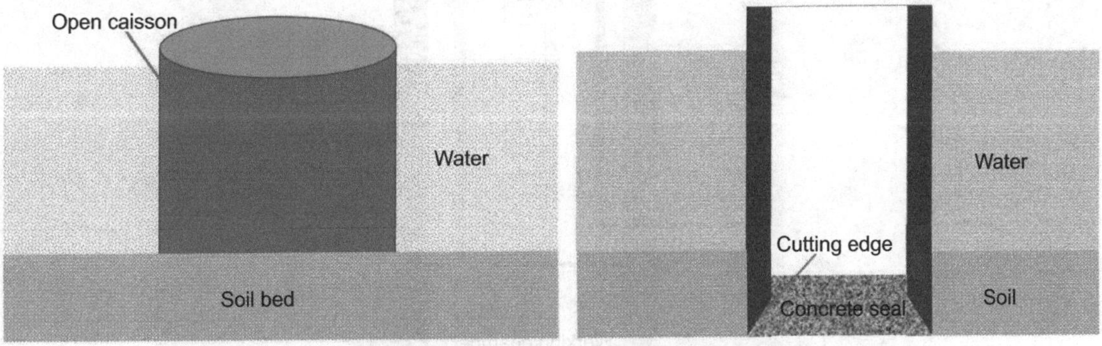

Fig. 2.22: Open caissons

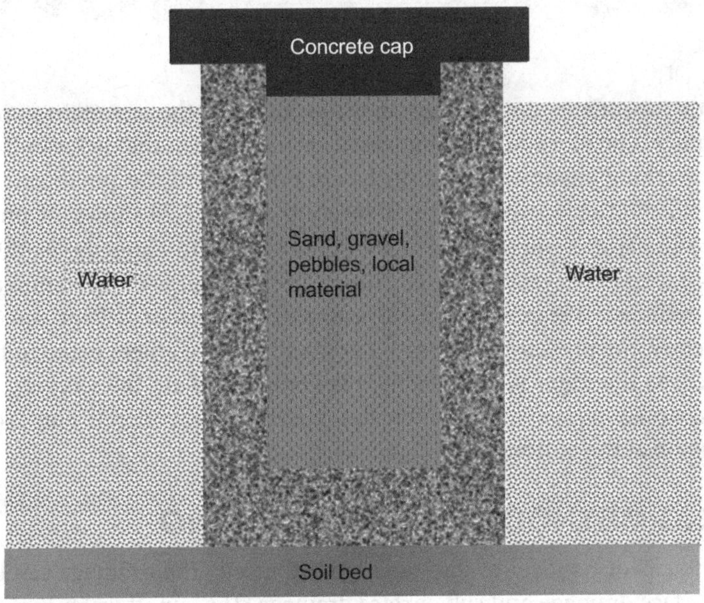

Fig. 2.23: Floating caissons

The floating caissons are closed at the bottom portion while they are open at the top. The open caissons are light in weight and are constructed on the ground. These are also hollow in nature. The floating caissons are floated to the destination where these are to be installed. The floating caissons are sunk into the underwater soil bed by increasing their weight. The weight of the floating caisson is increased by filling it with sand, rocks, concrete, etc. The floating caissons are not penetrated in the ground but are sat on the level surface. These are stable because of their weight. Figure 2.23 shows the floating caissons.

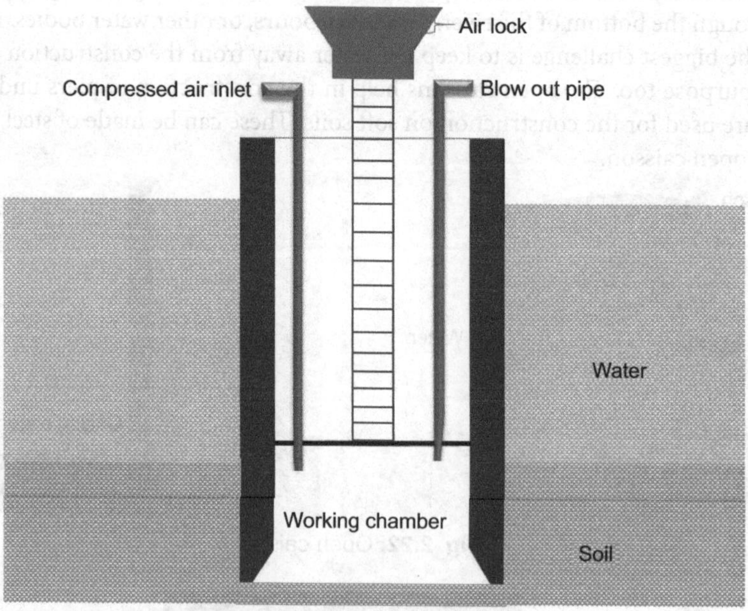

Fig. 2.24: Pneumatic caisson

The pneumatic caissons are constructed on the ground and these contain some working chambers. The working chamber is facilitated by a compressed air mechanism that helps in the prevention of entry of water into the working chamber. The airlocks are provided in the pneumatic caissons. Pneumatic caissons are very difficult to place at the desired location. Figure 2.24 shows the pneumatic caisson.

There are two shafts in a pneumatic caisson. One is used to enter the working chamber and the other is used to extract the excavated material. The compressed air is induced into the working chamber and the water is excluded. The pressure at the entry of the system is kept equal to the atmosphere. As the labourers enter, the pressure is increased gradually. The workers come to the working chamber through the ladder. As the water is excluded by the compressed air, the workers can excavate in a dry environment. As the caisson is penetrated at the desired depth, the concrete plug is constructed that seals the pneumatic caisson at the bottom.

2.9 PUMPING METHODS FOR DEWATERING

Drainage is the process of removal of the water from the soil. The drainage can be classified into two types which are surface drainage and sub-surface drainage. The runoff water that moves on the ground is diverged in any other direction to save the site from the flow of water. This type of drainage is known as surface drainage. If the water which is held in the pores of the soil is removed, this process is known as sub-surface drainage or dewatering.

In some places where excavation of trench is deeper and ground water table is high, the problem of water logging occurs. Sometimes the seeping water is very fast, it becomes impossible to carry excavation of the foundation. Therefore it becomes necessary to dewater the excavation area and keep it dry. The water table should be kept 0.5 m below the excavation of the foundation to keep the area dry. Several methods are available for dewatering and lowering the water table. Construction site information and types of soil are useful in deciding the economical and most suitable method for dewatering. The dewatering also improves the properties of the soil. The dewatering can be performed in two ways.

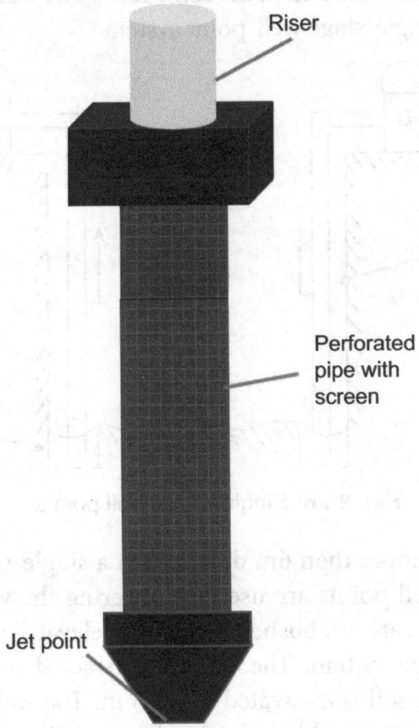

Fig. 2.25: Well point

If the water is removed only at the time of construction and the water level is maintained again, then it is called temporary dewatering. If the water is removed permanently and the water level is varied, it is called permanent dewatering. The pumps are used to extract the water from the soil mass.

2.9.1 Well Points

Well points are a pumping system to extract the water. A well point is a perforated (with small holes throughout the perimeter) pipe. The diameter of a well point is 5 cm and it is 1 m long. The holes on the surface of the well points are covered with a screen which prevents the blockage of the perforations. The bottom of the well point is provided with a conical steel drive. It helps in the installation of the well points. There is a jetting system at the bottom of the well point. The jetting system facilitates a strong jet of water which helps in the penetration of the well points. A ball valve at the bottom of the well point helps the water jet to be in the downward direction.

SCAN ME

The well points are connected to a pipe called riser pipe which is attached to a horizontal pipe of 15 cm to 30 cm in diameter. Figure 2.25 shows the well point.

The well points are installed and connected to the horizontal pipe. The horizontal pipe is connected to a pump. The pump sucks the water through the well points. A depression is formed around every well point. These cones of depression form a drawdown curve. The screening provided at the periphery of the well point can prevent the coarse material. So, the well points are suitable for soils with good permeability. These can lower the water table by 5 m to 6 m. In the case of shallow foundations, the well points are installed at a distance of 1 m to 3 m at the same level. The water is extracted as shown in figure 2.26. This system is known as a single-stage well-point system.

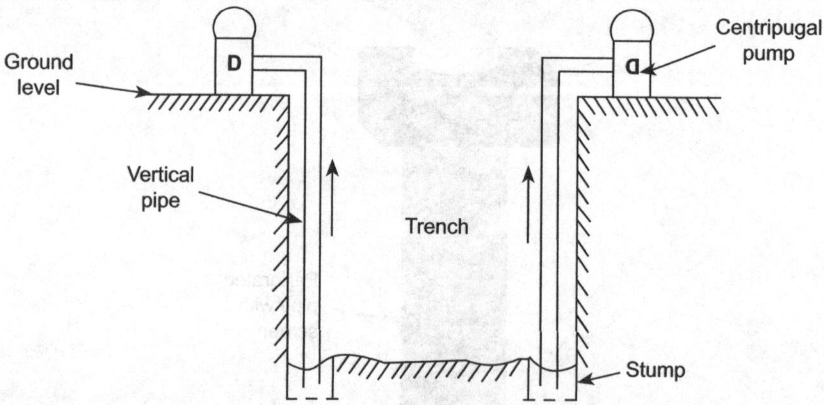

Fig. 2.26: Single-stage well points

If the water to be lowered is more than 6m depth, then a single-stage well point system can't give desired results. So, multi-stage well points are used for lowering the water table. In a multi-stage well point system, a series of well points are established at different elevations. The well points are established at stage 1 same as the single-stage system. The water is extracted up to 5 m depth. The excavation operations are conducted and the soil is excavated up to 5 m. The well point system are installed at a new depth now at stage 2 and the water table is further lowered. The process can be continued for the excavation of 15 m depth. If the water is to be removed beyond 15 m, deep well systems can be used. Figure 2.27 shows the multi-stage well-point system.

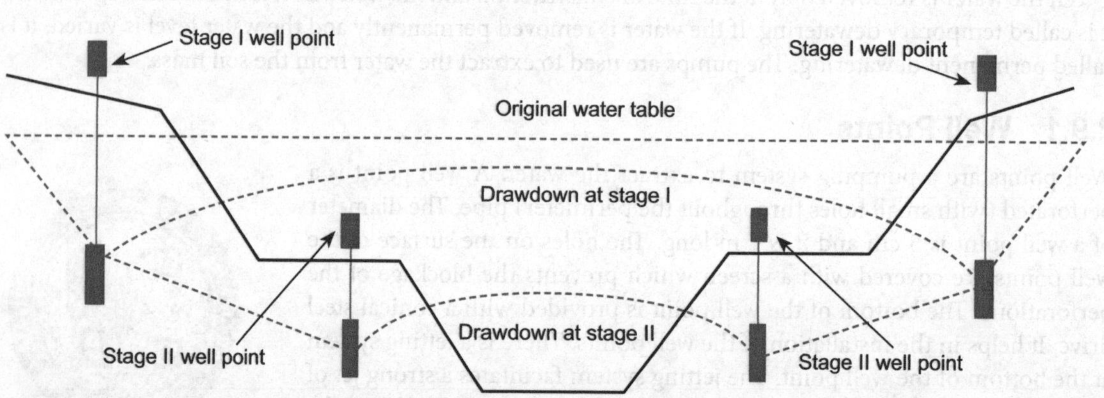

Fig. 2.27: Multi-stage well points

2.9.2 Vacuum Well Points

If the size of the soil particles is lesser than 0.05 mm, the normal well points can't be used. So, the silt, sand, and clays can't be dewatered with normal well points. In this case, a vacuum well point can be a good alternative.

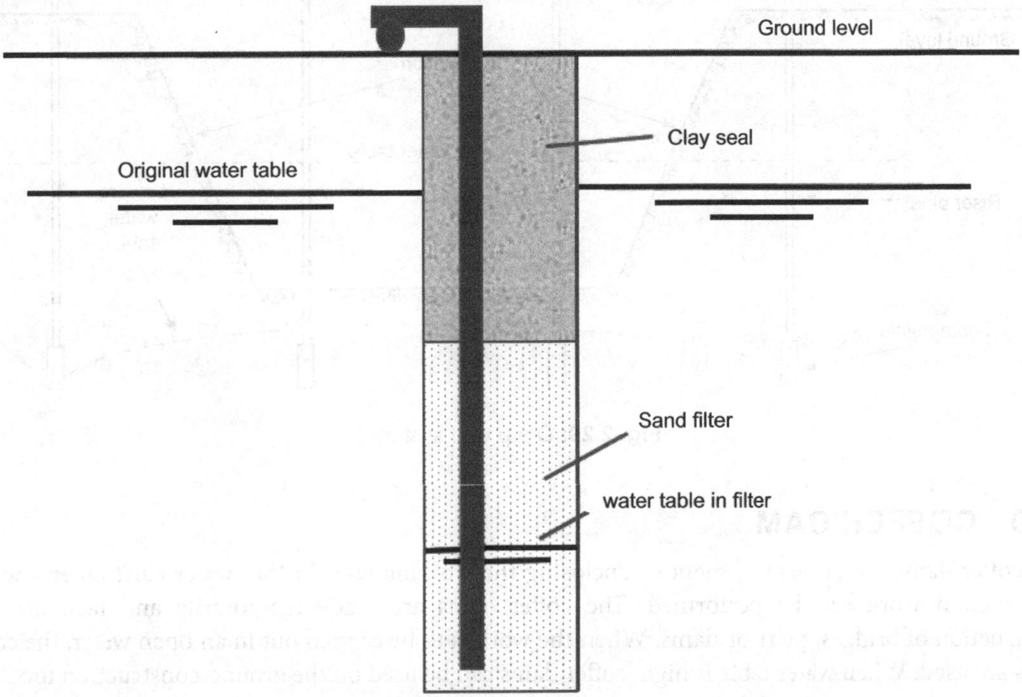

Fig. 2.28: Vacuum well points

The soils with low permeability are drained with vacuum well points. A hole of diameter 25 cm is bored around the well point and filled with coarse sand. The coarse sand is filled from the bottom leaving a 1 m portion from the top empty. This one meter is filled with stiff clay that acts as a seal. The other materials which are impervious in nature can also be used as a seal. The header pipe is connected to the vacuum point, a vacuum is formed. The hydraulic gradient is increased and the process of dewatering takes place. Figure 2.28 shows the vacuum well point.

2.9.3 Deep well system

The deep foundations requires dewatering up to larger depths. In this case, a deep well system is used for the purpose of dewatering. The diameter of a deep well varies from 30 cm to 60 cm. The deep wells are bored up to 15 m to 30 m depth. The filters as in case of vacuum pump can be provided depending upon the soil conditions. The submersible pump is placed at the bottom and protected by a casing. The space between deep wells can be 10 m to 30 m. A motor is attached to the pump which helps in deriving the pump. The deep wells are placed at the outer periphery of the region where the excavation is to be conducted. Figure 2.29 shows the deep well system.

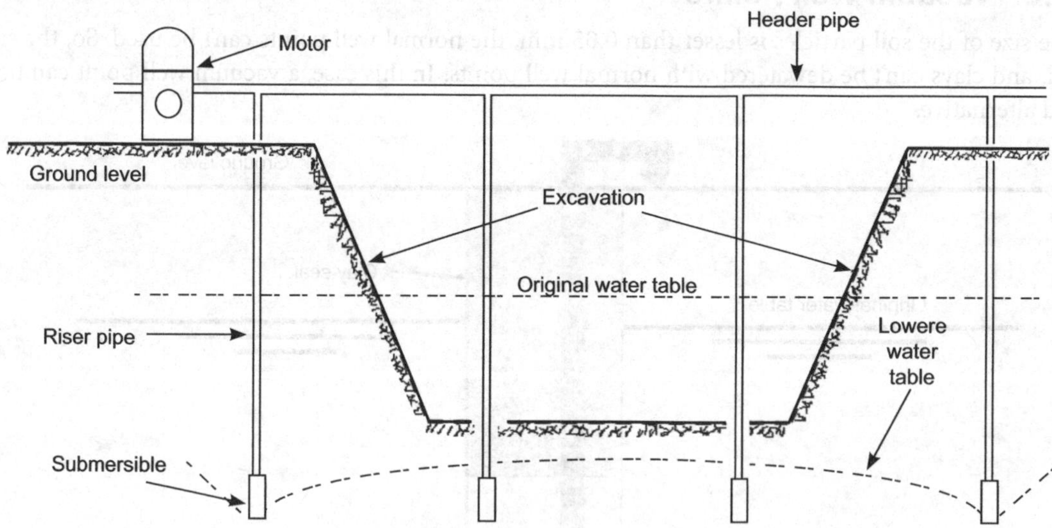

Fig. 2.29: Deep well system

2.10 COFFER DAM

The coffer dams are an arrangement of enclosing the working area so that water can't enter and the construction work can be performed. The coffer dams are made temporarily and facilitates the construction of bridges, piers or dams. When the work is to be carried out in an open water, the coffer dams are used. When water table is high, coffer dams can be used on the ground construction too.

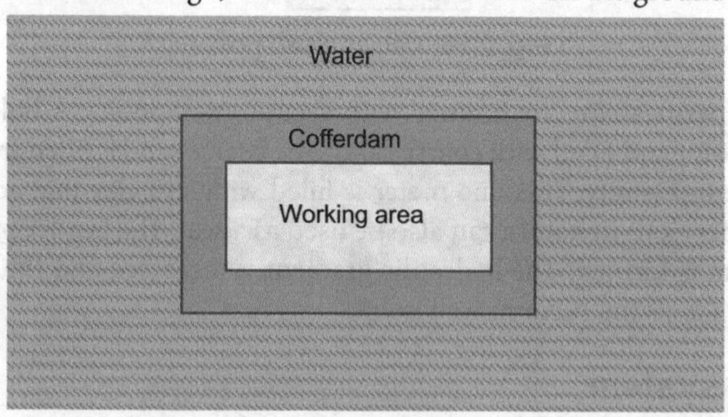

Fig. 2.30: Cofferdam

The water is stopped coming to the working area by the walls constructed or installed around the periphery of the working area. The cofferdams provide the dry working area where the construction work can be carried out. The cofferdams should be stable against the water currents.

The coffer dams can be constructed from the earth, sheet piles, rocks etc. Figure 2.30 shows a typical concept plan of the coffer dam.

The cofferdams can be of different types. The earth cofferdams form a barrier in the form of an embankment and restricts the entry of water to the working area. These are the simplest form of the cofferdams. The earth cofferdams are only suitable for the water depth of 3 m or less. The freeboard (height of the dam above water level) should be at least 1 m. Figure 2.31 shows the earth coffer dam.

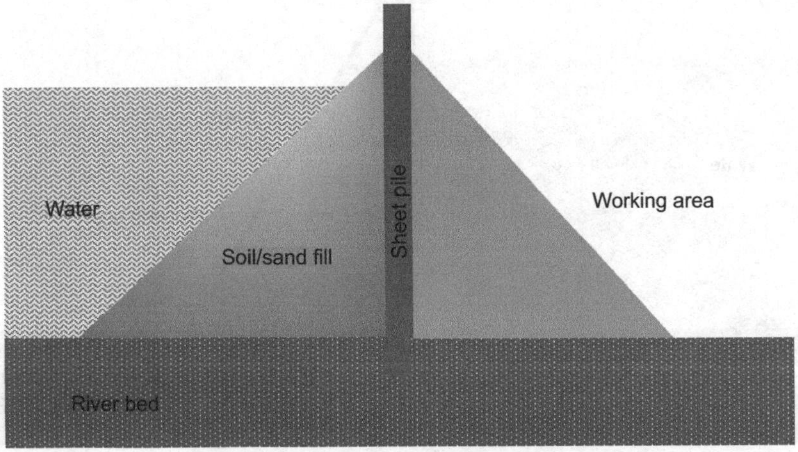

Fig. 2.31: Earth cofferdam

The rockfill cofferdam is constructed using the rocks laying in the groups. The rocks and pebbles used in the rockfill dam are pervious in nature and they can absorb the water. So, an impermeable membrane is provided at the surface of the rockfill cofferdam. This type of cofferdam remains stable due to its self-weight like the earth cofferdam. The rockfill cofferdam have an inclination of 1 in horizontal to 1.5 in vertical. It is not necessary that the cofferdams will be completely watertight structure. Some seepage is allowed due to economic consideration as making a completely water tight structure will be very costly. Figure 2.32 shows the rockfill cofferdam.

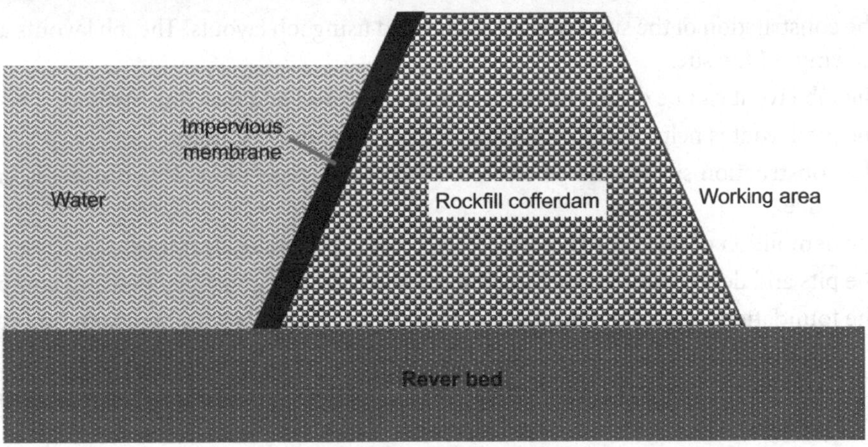

Fig. 2.32: Rockfill cofferdam

The water can be prevented by using the sheet piles. The single sheet pile coffer dam can be used for small sites. The sheet piles are connected and supported by strut from other side. The sheet pile cofferdams are suitable for 4 m depth. The velocity of the water should not be very high in case of sheet piles. Figure 2.33 shows the sheet pile cofferdam.

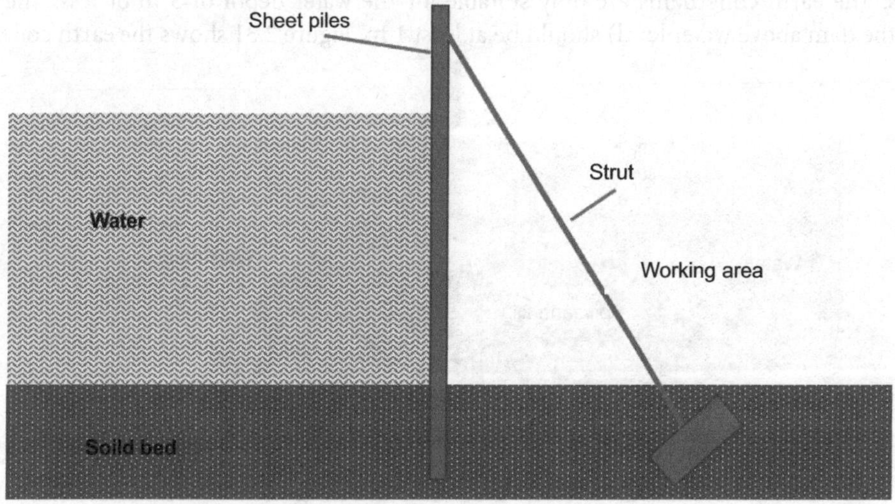

Fig. 2.33: Sheet pile cofferdam

The two-sided coffer dams can also be used. The sheet piles must be sealed properly for the construction of a cofferdam. There may be sheet piles of special shapes that can be used for the construction of a cofferdam. These specially shaped cofferdams are called cellular cofferdams. The cellular cofferdams are easily fitted in each other to make a watertight structure.

UNIT SUMMARY

- The construction of the substructure is planned using job layouts. The job layouts are plans and drawings of the site.
- The job layout can be drawn by the centre line method and face line method.
- The job layout is helpful in the management of time and resources.
- The construction site should be cleared of trees and vegetation. This process is known as site clearance.
- The demolition of the old structure if exists come under site clearance.
- The pits and depressions created during the site clearance must be filled and compacted.
- The foundation is excavated to the required depth. The soils with high standing capacity don't require any support.
- If the soil hasn't enough stability and strength, supports in the form of timbering and strutting are provided.
- The excavation of the foundation can be done manually or mechanically.
- Hoe, bulldozers, power shovels, tractors, etc. are the excavation machinery that facilitates the different requirements of excavation.
- The foundation can be classified into shallow and deep foundations.
- A shallow foundation has a depth equal to its width.
- Spread footings spread the load of the building components and transfer it to the earth.
- The pad column footing consists of an enlarged concrete bed under the base of the column.

- The stepped footing transfers the load gradually in steps. The base is enlarged by masonry and a concrete bed is provided under the steps.
- Combined footing consists of a platform that supports more than one column. If the load of the two columns is equal, a rectangular combined footing is provided otherwise a trapezoidal footing is provided.
- A grillage footing is constructed of steel or timber beams and can take heavy loads.
- A raft or mat foundation covers the whole area under the building and supports all the components of the building.
- The raft foundation transfers the load by slab action and reduces the differential settlements.
- When the hard stratum lies under larger depth, the deep foundations are used.
- The piles can be made of concrete, timber or steel. These transfer the load with the help of their bottom tip or by the skin friction or by a combined action.
- Caissons are the foundation which are provided under the water.
- The pneumatic caissons consist of working chambers which remain free from water under the water channels.
- The pumps are used for dewatering the construction area.
- The cofferdams surround the working area and prevents the entry of the water to help the construction activities.

EXERCISES

Multiple Choice Questions

1. What does a job layout represent?
   (b) Building plan (b) Components of building
   (c) Site plan (d) All of the above
2. Which of the following is an essential part of a job layout?
   (a) Entry point (b) Exit point
   (c) Storage area (d) All of the above
3. Which of the following is not a method of making a job layout?
   (b) Critical path method (b) Centre line method
   (c) Face line method (d) None
4. The excavation is supported by:
   (a) Grouting (b) Underpinning
   (c) Timbering and Strutting (d) None
5. The foundation which connects the columns with a beam:
   (b) Stepped footing (b) Strap footing
   (c) Grillage footing (d) Strip footing
6. For a lightly loaded small structure, which foundation will you use?
   (b) Grillage foundation (b) Mat footing
   (c) Strap footing (d) Strip footing

7. During a site visit, I saw steel stanchions supported on a combination of steel beams. Which type of foundation is this?
   (b) Mat foundation (b) Pier foundation
   (c) Grillage foundation (d) None of the above
8. Pier foundation is also known as:
   (b) Box foundation (b) Bridge foundation
   (c) Caisson foundation (d) None
9. If the hard stratum is not available even after 15 m depth, which type of foundation is recommended?
   (b) End bearing pile (b) Friction piles
   (c) Pier foundation (d) All of the above
10. In the vacuum well point, the top of the well is sealed by:
   (b) Bentonite (b) Clay
   (c) Soil cement (d) All of the above
11. The special shape of sheet piles is used in which type of cofferdam?
   (a) Cellular cofferdam (b) Single sheet pile cofferdam
   (c) Braced cofferdam (d) None
12. The earth-fill cofferdams can be used only up to the depth of water:
   (a) 1 m (b) 7 m
   (c) 3 m (d) More than 10 m

Answers of Multiple Choice Questions

1. (c), 2. (d), 3. (a), 4. (c), 5. (b), 6. (d), 7. (c), 8. (c), 9. (b), 10. (d), 11. (a), 12. (c)

Short and Long Answer Type Questions

1. Explain the difference between shallow and deep foundation
2. Write a short note on raft footing.
3. Enlist the conditions in which combined footing is provided.
4. What type of foundation can be provided for the structure with light loading? Discuss in detail.
5. A region has weak soil and faces high differential settlement. Identify the type of shallow foundation to be provided for this region.
6. Explain the difference between spread and stepped footing
7. Explain different hand tools used for excavation.
8. What do you understand by timbering and strutting? Explain the different methods with diagrams.
9. What are the different precautions taken at the time of job layout?
10. Explain the rock-fill coffer dams.
11. Draw a neat sketch of earth fill coffer dam. Identify its various parts.
12. Explain the different materials for filling of the plinth.

13. Explain centre-line method for the layout.
14. Identify the suitability of different types of piles based on their material.
15. In what type of situations, a caisson foundation can be provided?
16. What are the different types of caissons?
17. Explain the working of pneumatic caissons.
18. Write a short note on open caisson.
19. Write a short note in the vacuum well point.
20. Draw a neat sketch of multi-stage well points. Identify its various components.
21. Explain different tools used for deep foundation excavation.
22. Explore the various manual tools of excavation.

KNOW MORE

Taj Mahal is one of the finest architectural marvel in the world. The monument is famous due to its symmetrical design and aesthetics. The construction of Taj Mahal started in 1632 and completed into 1653. Even after centuries, the monument is standing intact. Do you know what type of foundation is provided in this monument? The Taj Mahal is built over a well foundation. To know more about the foundation of Taj Mahal, scan the bar code.

SCAN ME

REFERENCES AND SUGGESTED READINGS

1. Chudley, R., Greeno, R. *Building Construction Handbook.* Elsevier, Butterworth-Heinemann, Amsterdam, 2007.
2. Punmia, B. C., et al. *Building Construction: An Elementary As Well As Advanced for Engineering Students.* Laxmi Publications, 2016.
3. Iano, J., Allen, E. *Fundamentals of Building Construction: Materials and Methods.* Wiley. 2019.
4. Arora, K. R. *Soil Mechanics and Foundation Engineering.* Standard Publishers, 2019.

3 Construction of Superstructure

UNIT SPECIFICS

Through this unit we have discussed the following aspects:

- To understand the types of stone masonry and their suitability
- Process of selection of stone masonry
- Overview of brick masonry and its requirements
- To understand the different types of bonds in bricks
- Discussion on composite masonry
- To understand the different types of scaffolding and shoring
- Overview of the process of removal of the formwork

For helping the students in the visualization of the content and to enhance their practical knowledge, the figures are provided in 2D and 3D. The content given in this unit provides practical knowledge and will help the students in decision-making during the fieldwork.

This unit contains short and long answer-type questions along with multiple choice questions, a list of references, and suggested reading is given so that one can go through them for practice. Some QR codes are given which can be scanned for more information on various topics of interest and to get extra knowledge. Some interesting facts are given under 'Know more'.

RATIONALE

The unit on the construction of superstructure helps the students to develop an idea about different types of masonry used for the construction. Knowledge about the different types of stone masonry and brick masonry is provided. The suitability of different types of masonry is also discussed in this unit. The students get an idea about the junctions in brick masonry and joints in stone masonry. The students will be able to understand the process of erection of scaffolding and shoring. The students will get knowledge about precautions to be taken during the removal of formwork.

The knowledge of masonry works is essential for the students. The students can apply the knowledge of the construction of superstructures during their fieldwork. Knowledge gained in this unit will help the students in the selection of suitable masonry for different conditions. The students can apply the knowledge in the selection and implementation of suitable shoring or scaffolding technique during the construction work. The concepts in this unit will help the student in effective supervision of construction works of the superstructure.

PRE-REQUISITES

Basic knowledge of excavation and foundation is required.

UNIT OUTCOMES

The list of outcomes of this unit is as follows:

U3-O1: To explain the different types of stone and brick masonry
U3-O2: To understand the suitability of different types of masonry
U3-O3: To understand the different types of bonds in brick masonry
U3-O4: To select the suitable type of scaffolding and shoring
U3-O5: To understand the precautions to be taken during the construction of the superstructure

Unit-3 Outcome	EXPECTED MAPPING WITH COURSE OUTCOMES (1 - Weak Correlation; 2 - Medium Correlation; 3 - Strong Correlation)				
	CO1	CO2	CO3	CO4	CO5
U3-O1	1	-	2	-	-
U3-O2	-	-	3	-	-
U3-O3	-	-	2	-	-
U3-O4	-	-	1	1	-
U3-O5	-	-	-	1	-

3.1 INTRODUCTION

In the previous unit, you learned about the construction of the substructure. In this unit, we will learn about the construction of the superstructure. The superstructure is constructed by bonding the masonry units with mortar. The masonry units can be stones in the case of stone masonry or bricks in the case of brick masonry. Masonry can be used for the construction of foundations, walls, columns, and other building components. The masonry in a building plays multiple roles. It gives support to the loads. It divides the space in the building. The masonry provides insulation and soundproofing to the building. It also saves the building from the impact of weather and fire. Stone masonry and brick masonry are used in construction extensively.

3.2 STONE MASONRY

Stone masonry consists of mortar and stones. Mortar is a paste of binding material, some inert material like sand and water. Mortar helps in binding the stones. The mortar can be mud or clay that can bind the stones. The mortar can be produced by a combination of cement, sand, and aggregates. Lime can also be used as binding material. Lime mortars can be prepared with hydraulic or semi-hydraulic lime. The mortars should be prepared by properly mixing their components and tested for compression.

SCAN ME

Stones are used in construction widely. The stone should have enough strength to bear the superimposed loads. It should remain unaffected by the impact of weathering and the environment. The stones are obtained by blasting the rocks into pieces. The rocks are classified into the following three categories:

(1) Igneous rocks
(2) Sedimentary rocks
(3) Metamorphic rocks

The material under the earth's surface remains under intense pressure and heat. Due to the impact of pressure and heat, the rocks and other materials remain in a molten form which is known as magma. The igneous rocks are made due to the solidification of the magma. If the magma solidifies under the surface of the earth, it is known as intrusive igneous rock. If the magma comes out of fissures and cracks and solidifies above the surface of the earth, it is known as extrusive igneous rocks. Granite, pegmatite, pumice, etc. are examples of igneous rocks.

The sedimentary rocks are formed from the fragments of the other rocks. The rocks are broken into small fragments due to the action of water, wind, and erosion. These fragments are transported by wind and water and are deposited at some place. As time goes on, the fragments are deposited to form sedimentary rocks. Shale, chalk, and sandstone are examples of sedimentary rocks.

Due to intense pressure and heat, the natural rocks are changed in their nature. These types of rocks are known as metamorphic rocks. Metamorphic rocks can be divided into foliated and non-foliated categories. In foliated rocks, the elongated particles align in the perpendicular direction of the pressure while in non-foliated rocks, the alignment of the particles remains in the line of the pressure. Gneiss is an example of a foliated metamorphic rock while marble is a non-foliated metamorphic rock.

The engineers should be able to identify the type of rocks by their appearance. If you want to see how different rocks appear, you can scan the bar code given in this section.

3.3 TERMS USED IN STONE MASONRY

The terms used frequently in the case of stone masonry are discussed as follows:

(1) **Face and facing:** The surface which is on the outer side and is exposed to external agencies like weather, wind, rainfall, sunlight, etc. is known as the face. The material used on the face is known as facing.

(2) **Back and backing:** The inner surface of the wall which is not exposed to external agencies is known as the back and the material used in the back is known as the backing.

(3) **Hearting:** Hearting in stone masonry may be defined as the space of wall between facing and the backing. It is also known as hearting and filling.

(4) **Through stone:** A through stone may be defined as the stone placed perpendicular to the length of the wall. If the thickness of the wall is less, the length of the through stone may be equal to the width of the wall. For the thick walls, multiple through stones can be used.

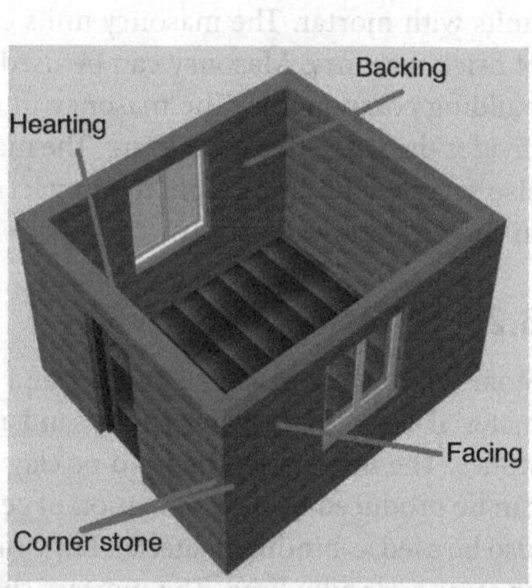

Fig. 3.1: Terms used in stone masonry

(5) **Cornerstone:** Stone that is used at the corner of the wall when two walls meet at the right angle. Figure 3.1 shows the details of the terms used in stone masonry.

(6) **Cornice:** A cornice may be defined as a projecting course at the joint of the ceiling and wall to dispose of the rainwater. It is made aesthetically pleasing and throated. The cornice is provided with a weight of parapet on it to avoid its overturning, Cornice runs through the width of the wall.

(7) **Bed:** It is the bottom of the stone at which it rests in a course or line.

3.4 TYPES OF STONE MASONRY

Stone masonry can be classified based on the arrangement and finishing of the stones. Stone masonry can be classified into the following two categories:

- Rubble masonry
- Ashlar masonry

3.4.1 Rubble Masonry

The rubble masonry is also known as rubblework. The stones used in the rubble masonry are undressed and rough. The stones which don't go through the process of dressing for surface finishing are used in rubble masonry. Dressing of the stones may be defined as the process of giving the stones a required shape and size. Due to the irregular size of the stones, the joints in the rubble masonry are wide comparatively. The dressing of the stones is a very difficult task. So, the rubble masonry can be used in the hidden layers like the course of the foundation. The different types of rubble masonry are classified into the following categories:

- Dry rubble masonry
- Random rubble masonry
- Square rubble masonry
- Polygonal walling masonry
- Flint walling

3.4.1.1 Dry Rubble Masonry

The simplest type of the rubble masonry is the dry rubble masonry. The stones of different shapes and dimensions are placed one over other without mortar. The stones remain stable due to the friction and gravity. The stones can be shaped using chisel and hammer. The dry rubble stone masonry can be used for the formation of the boundary walls and retaining walls as shown in the figure 3.2. The retaining walls formed by laying the stones without the mortar are known as gabion walls. The setting of the stones requires skills so that the stones don't move and are placed properly.

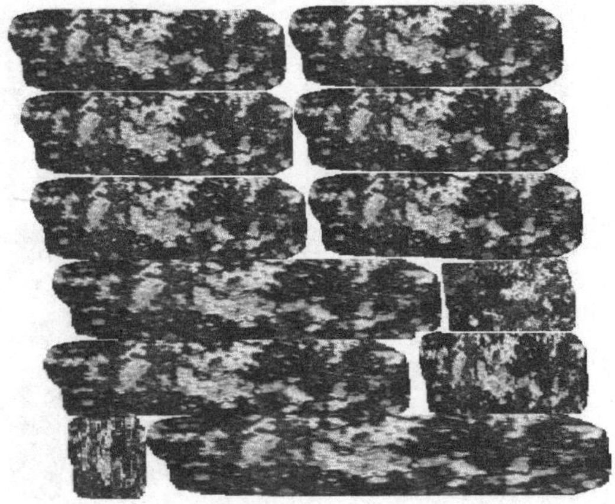

Fig. 3.2: Dry rubble masonry

3.4.1.2 Random Rubble Masonry

The unfinished and rough stones directly from the quarry are used in random rubble masonry. The random rubble masonry can be coursed or uncoursed. The uncoursed random rubble masonry is the roughest form of the stones. The stones used in this type of masonry are of different sizes. The stones with larger dimensions are used at the corners. The arrangement of the stones is adjusted based on the different sizes. The courses or lines of the stones are not maintained in this type of masonry. The lime mortar or cement mortar can be used for joining the stones. During the construction of random rubble masonry, continuous vertical joints are avoided. Figure 3.3 (a) shows the uncoursed random rubble masonry.

Fig. 3.3 (a): Uncoursed random rubble

The random rubble masonry can also be used in the coursed form. The courses or lines are maintained in this type of construction. The stones at the edges are laid first and the space between these two stones is filled with a combination of different-sized stones for roughly maintaining a course. Figure 3.3 (b) shows the coursed random rubble masonry. The series of stones represent the way construction proceeds. The stones Q1 and Q2 are placed in the first step and the stones B1 to B4 are placed in the second step. The stones from A1 to A5 are placed the last.

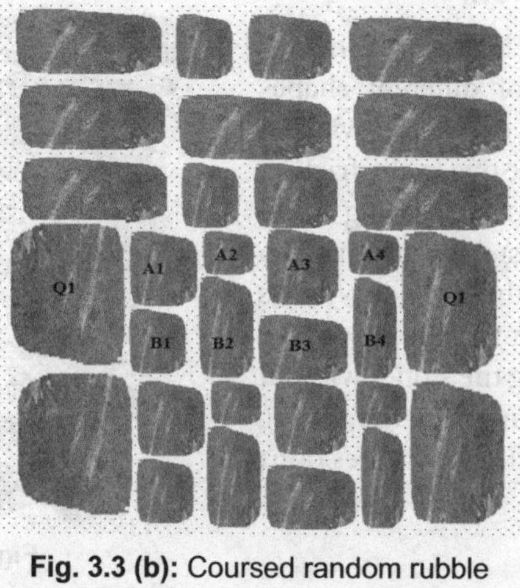

Fig. 3.3 (b): Coursed random rubble

3.4.1.3 Square Rubble Masonry

The square rubble masonry consists of stones that are straightened on the edges. The stones are squared using the hammer and chisel. So, the appearance of the square rubble masonry is better as compared to the random rubble masonry. The square rubble masonry can be arranged in three types. In uncoursed squared rubble masonry, stones of varying sizes are used.

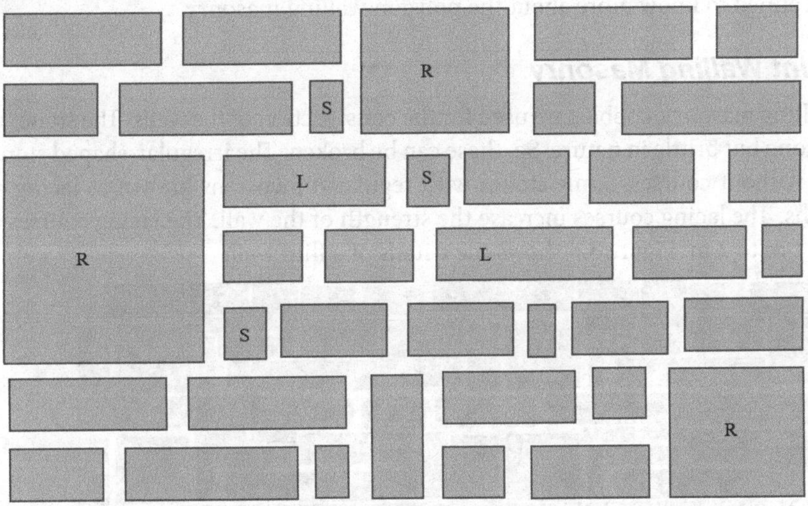

Fig. 3.4: Uncoursed square rubble masonry

The appearance in the uncoursed square rubble masonry can be improved using different-sized stones. A riser is a through stone that is bigger in size. A leveler is a thin stone. Sneck and check are the small-sized stones. So, a pattern of the riser, leveler, and sneck-check stones can be established for a better appearance. Figure 3.4 shows the uncoursed square rubble masonry. R represents the riser, L represents the leveler and S represents the sneck and checks stones.

The square rubble masonry can be arranged in the courses too. In the case of coursed square rubble masonry, irregular courses can be provided. The height of the courses may vary but the stones are arranged in lines. This is known as built-in course square rubble masonry. It is known as regular coursed square rubble masonry if the courses have the same height. Figure 3.5 (a) shows irregularly coursed square rubble masonry while 3.5 (b) shows the regularly coursed square rubble masonry.

Fig. 3.5 (a): Irregular coursed square rubble

Fig. 3.5 (b): Regularly coursed square rubble

3.4.1.4 Polygon Walling Masonry

The stones with irregular polygon shapes are used in the polygon walling masonry. The stones are placed in a way to fit each other. The stones used in polygon walling can be well dressed and should fit each other by varying their shapes. Polygon walling is still used in in the country like Greece. The QR code can be scanned to know more about the polygon walling masonry.

3.4.1.5 Flint Walling Masonry

In the flint walling masonry, cobbles are used for the construction of the walls. The stones used in the flint walling are strong but brittle in nature. So, these can be broken. The irregular-shaped stones are arranged in courses or without courses. Some stones with regular dimensions known as lacing are provided at different depths. The lacing courses increase the strength of the wall. The lacing courses are provided at an interval of 1 m to 2 m. Figure 3.6 shows the details of a flint wall.

Fig. 3.6: Details of flint wall

3.4.2 Ashlar Masonry

The ashlar masonry consists of stones that are dressed finely and have the same shape and size. The stones of ashlar masonry require significant effort and time for their dressing and shaping. The joints in the ashlar masonry remain uniform in size and are thin as compared to the rubble masonry. The heights of the stones remain 25 cm to 30 cm. The stones of the same height are kept in a course but the height of the courses may vary. The appearance of the ashlar masonry is better than the rubble masonry. Due to the thin joints, the strength and stability of ashlar masonry are very high. It can be classified into the following six types:

- Ashlar rough tooled masonry
- Ashlar fine tooled masonry
- Ashlar rock-faced masonry
- Ashlar chamfered masonry
- Ashlar coursed masonry
- Ashlar facing masonry

UNDERGRADUATE COURSE BOOKS

S.No.	ISBN	Book Name	Author
38	9789355385130	Hydraulic Engineering	Sayantan Ganguly
39	9789355385550	Hydrology & Water Resources Engineering	Mitthan Lal Kansal, P.K. Agarwal
40	9789355389688	Information Technology Workshop in MATLAB	Hari Prabhat Gupta
41	9789355383235	Instrumentation for Civil Engineering Applications	Ashim Kanti Dey, Manita Das
42	9789355387042	Introduction to Civil Engineering	Manabendra Saharia, Nagendra R. Velaga
43	9789355384805	Introduction to Database Systems	Vikram Singh
44	9789355387509	Machine Element & System Design	Sabah Khan
45	9789355385383	Machine Learning	Himanshu Mittal
46	9789355388414	Manufacturing Automation	Chitragupt Swaroop Chitransh
47	9789355387233	Manufacturing Processes	Dheerendra Kumar Dwivedi
48	9789355389886	Materials Engineering	Dheerendra Kumar Dwivedi
49	9789391505172	Mathematics - I Calculus and Linear Algebra [For Non-Computer Science Engineering Branches]	Reena Garg
50	9789391505271	Mathematics - I Calculus and Linear Algebra [For Computer Science Engineering Branches]	Reena Garg
51	9789355384607	Measurements & Metrology	V.P. Chandramohan
52	9789355383730	Mechanics of Materials	Arvind Vyavahare
53	9789355383358	Mechatronics, Robotics & Control	J. Srinivas
54	9789355385116	Microprocessor and Microcontroller	Saurabh Chaudhury, Risha Mal
55	9789355386083	Microprocessors	Yogendra Kumar Gupta
56	9789355389169	Mobile Communication and Networks	Vimal Bhatia
57	9789355385475	Operating Systems	Sukomal Pal
58	9789355385604	Operating Systems	Sanjaya Kumar Panda
59	9789391505134	Physics (Oscillations, Waves and Optics) With lab manual	A.B. Bhattacharya, Atanu Nag
60	9789391505257	Physics (Quantum Mechanics for Engineers) With lab manual	A.B. Bhattacharya, Atanu Nag
61	9789391505165	Physics (Introduction to Electromagnetic Theory) (with Lab Manual)	A.B. Bhattacharya, Atanu Nag
62	9789391505059	Physics (Introduction to Mechanics) (with Lab Manual)	A.B. Bhattacharya, Atanu Nag
63	9789355385956	Power Electronics: Theory and Practicals	Lalit Chandra Saikia
64	9789355384942	Power Systems-I	Suresh Mikkili
65	9789355386236	Power Systems-II	Arun Rahul S, Manas K. Jena
66	9789355383754	Product Innovation & Entrepreneurship	Vigneswaran C
67	9789355384584	Production & Operation Management	Navin Kumar Dev
68	9789391505219	Programming for Problem Solving (with Lab Manual)	R.S. Salaria
69	9789355385451	Signals and Systems	Sanjay L. Nalbalwar
70	9789355386021	Strength of Materials	Uday Shanker Dixit, Nelson Muthu, S. M. Kamal
71	9789355383792	Structural Engineering	T.D. Gunneswara Rao, D. Ravi Prasad
72	9789355387004	Surveying and Geomatics	P. K. Garg
73	9789355384317	Testing & Evaluation of Civil Engineering Materials	Sandeep Chaudhary, Kashyap A Patel
74	9789355382306	Theory of Computation	Debi Prasanna Acharjya
75	9789355384928	Transportation Engineering	A.K. Raji
76	9789355385208	VLSI Design	Mahendra Sakare, Shonal Chouksey
77	9789391505332	Workshop / Manufacturing Practices (with Lab Manual)	Veeranna D.K.

DIPLOMA COURSE BOOKS

S.No.	ISBN	Book Name	Author
1	9789355382481	Aanandam An Act of Community Service	Shuchi Sharma
2	9789355385567	Advanced Manufacturing Processes	Ranjeet Kumar Sahu, Devendra Laxman Kamble
3	9789355385109	Advanced Surveying: Theory & Practice	Ramakant Agrawal, Parshottam Sarathe
4	9789355384546	Algorithms	Piyoosh P, Arnab Sarkar
5	9789391505448	Applied Chemistry (with Lab Manual)	Anju Rawlley, Devdatta Vinayakrao Saraf
6	9789391505431	Applied Physics-I (with Lab Manual)	Vinod Kumar Yadav, Mina Talati
7	9789391505578	Applied Physics-II (with Lab Manual)	Hussain Jeevakhan
8	9789355387530	Basic Mechanical Engineering	Dharmendra Singh, Chandra Shekhar Rajoria
9	9789355387394	Building Construction	Rinku Kumar, Sandeep Panchal
10	9789355383280	Building Electrification	Arnab Ghosh
11	9789355384379	Building Planning and Drawing	Aparna Dwivedi
12	9789391505493	Communication Skills in English	Anjana Tiwari
13	9789355383570	Computer Aided Machine Drawing Practice	Kanak Kalita
14	9789355388964	Computer Networking and Data Communication	Sanjaya Shankar Tripathy
15	9789355382986	Computer Networks: Theory & Practicals	Brijendra Pratap Singh, Manoj Madhava Gore
16	9789355382672	Computer Programming: Theory and Practicals	Satyendra Singh Chouhan
17	9789355387714	Computer System Organization	Sonal Yadav
18	9789355382658	Concrete Technology: Theory and Practice	Sasankasekhar Mandal, Vishwajit Anand, Sushil K. Agarwala
19	9789355383334	Construction Materials and Testing	Vanita Aggarwal
20	9789355383501	Consumer Electronics	Amit M. Joshi, Maulin M. Joshi, Urvashi Prakash Shukla
21	9789355387684	Data Structures: Theory & Practicals	NB Venkateswarlu
22	9789355388513	Design of Machine Elements	A. Kumaravel, M. Kathirselvam
23	9789355388360	Design of Steel and RCC Structure	Sparsh Johari
24	9789355383303	Digital Communication	Muralidhar Kulkarni, K. S. Shivaprakasha
25	9789355387295	Digital Electronics	Menka Yadav
26	9789355389435	Electric Circuits and Networks	Jamuna K., Nilanjan Tewari
27	9789355388452	Electric Motors and Transformers (Theory and Practicals)	Sanjay B. Bodkhe
28	9789355385901	Electric Power Transmission and Distribution	Sudhir Sharma
29	9789355389466	Electrical & Electronics Measurements and Instrumentation	Sudarsan Sahoo, Vipin Chandra Pal, Sudipta Chakraborty
30	9789355382696	Electronic Communication	M. D. Selvaraj, P. Nagaradjane
31	9789355387202	Electronic Devices and Circuits	Ankesh Jain
32	9789355386373	Electronic Equipment Maintenance	Chanchal Sharma
33	9789355384690	Embedded Systems	Deepa M, G. Santhanamari, Susithra. N
34	9789355384867	Energy Conservation and Audit	M. Rizwan, Majid Jamil
35	9789391505509	Engineering Graphics	Sharad K. Pradhan, K.K. Jain
36	9789391505608	Engineering Mechanics (with Lab Manual)	Bhankhar Bharat, Gokaldas Vandana Somkuwar

DIPLOMA COURSE BOOKS

S.No.	ISBN	Book Name	Author
37	9789391505516	Engineering Workshop Practice	A. K. Sarathe
38	9789391505653	Environmental Science	Subrat Roy
39	9789355384966	Estimating, Costing and valuation	Sandeep Panchal
40	9789355389510	Fluid Mechanics & Hydraulic Machinery	Suman Chakraborty, Sourav Mitra, Aditya Bandopadhyay
41	9789391505592	Fundamentals of Electrical and Electronics Engineering (with Lab Manual)	Susan S. Mathew, Saji T. Chacko
42	9789355387240	Fundamentals of Power Electronics	Soumitra Kumar Mandal
43	9789355389602	Geotechnical Engineering (Theory & Practicals)	Neelima Satyam
44	9789355386137	Induction, Synchronous, and Special Electrical Machines	K. Selvajyothi, S. Tamilselvi, Sarat Kumar Sahoo
45	9789355382290	Industrial Engineering & Management	A. Manimaran
46	9789355384959	Internet of Things	Sujata Pal
47	9789355388650	Introduction to DBMS: Theory & Practicals	Myneni Madhu Bala
48	9789355384294	Introduction to e-Governance	Kamal Kant Verma
49	9789355386564	Introduction to Electric Generation Systems	Lalit Chandra Saikia, Nalin Behari, Deb Choudhury
50	9789391505585	Introduction to IT Systems (with Lab Manual)	Prashant Joshi
51	9789355382900	Linear Integrated Circuit	B. T. Krishna
52	9789355388148	Manufacturing Engineering	Santosh Kumar
53	9789355383877	Material Science & Engineering	Vinod Kumar, Nisheeth k. Prasad
54	9789391505424	Mathematics-I	Deepak Singh
55	9789391505523	Mathematics-II	Garima Singh
56	9789355385994	Measurements and Metrology	A. Gnanavelbabu, M. Santhana kumar, R. Saranraj
57	9789355388131	Mechanics of Materials and their Testing	R. Velmurugan, Pragathi
58	9789355385048	Microcontroller Applications	Ritula Thakur
59	9789355385062	Microcontrollers and Applications	Santanu Chattopadhyay
60	9789355388100	Mobile and Wireless Communication	Maulin Joshi, Urvashi P. Shukla
61	9789355387790	Production & Operations Management	Mayank Kumar Agrawal
62	9789355388704	Public Health Engineering	Rinku Kumar
63	9789355387455	Python Programming	Rupesh Nasre
64	9789355387189	Renewable Energy Technologies	Sanjay Agrawal, Rajeev K. Mishra
65	9789355387479	Software Engineering	Salman Abdul Moiz
66	9789355389749	Theory of Machines & Mechanisms	G. C. Mohan Kumar
67	9789355387332	Theory of Structures	Arunachalam Subramanian Balu
68	9789355389015	Thermal Engineering - I	V. Raghavan
69	9789355389831	Thermal Engineering - II	Parvesh Antil
70	9789355386779	Transportation Engineering (Theory & Practice)	A.K. Raji, K.K. Babu
71	9789355384287	Water Resource Engineering (Theory & Practice)	Prakash C. Swain, Dillip K. Ghose
72	9789355383051	Web Technologies: Theory & Practical	Raman Dugyala

3.4.2.1 Ashlar Rough Tooled Masonry

The sides and the bed of the stones used in the ashlar rough tooled masonry are dressed finely, but the stone's surface remains roughly dressed. Such type of ashlar masonry is known as bastard ashlar. The exposed face of the ashlar rough tooled masonry gives a harsh and rough appearance. The minimum thickness of the course in this type of masonry should be 15 cm. For a stone of edge length of 600 mm, the height of any point on the surface should not vary by more than 3 mm. The points on the bed should not vary by more than 3 mm with respect to the edges. For the side surface that forms the joints in ashlar rough tooled masonry, the variation of any point with respect to a straight edge should not be more than 6 mm. The edges are kept perfectly square. A strip on the perimeter of each stone is provided that may be approximately 25 mm wide and it is made by the chisel as shown in the figure 3.7.

Fig. 3.7: Ashlar rough tooled masonry

3.4.2.2 Ashlar Fine Tooled Masonry

The ashlar fine tooled masonry is the finest form of stone masonry. The stones used in this type of masonry are dressed very finely. The edges and surface are well dressed for straightness.

Fig. 3.8: Ashlar fine tooled masonry

Chisel is used to smoothen the surface and the surface is made even. The variation of any point with respect to the straight edge of 600 mm length should not be more than 1 mm in any direction. No point of the top and the bed should vary by 3 mm. Side surfaces that are responsible for the formation of joints should not vary by 6 mm. All the angles should be square and there should be no chippings on the edges. The thickness of the course should not be less than 15 cm. The maximum thickness of the mortar layer can be 5 mm. Figure 3.8 shows the ashlar fine tooled masonry.

3.4.2.3 Ashlar Rock Faced Masonry

In ashlar rock-faced masonry, the surface that is exposed to external agencies is not dressed and remains rough. A strip of around 25 mm around the perimeter is provided around each stone. Chisel is used in the case of ashlar rock-faced masonry. The edges and bottom of the stones are kept perfectly straight. The undulation or points above the surface of the stone should not extend beyond 75 mm. The projections above 75 mm can be trimmed. The height of the courses may vary from 15 cm to 30 cm. The ashlar rock-faced masonry looks like rough tooled masonry. The only difference is that the surface in the rough tooled masonry is dressed roughly while the surface is not dressed in the case of ashlar rock-faced masonry.

3.4.2.4 Ashlar Chamfered Masonry

Ashlar chamfered masonry is a special type of stone masonry in which the surface of the stones is bevelled around the periphery. The edges are bevelled at the angle of 45° with the help of a chisel. The depth of the groove is normally 25 mm. Figure 3.9 shows the ashlar chamfered masonry.

Fig. 3.9: Ashlar chamfered masonry

3.4.2.5 Ashlar Coursed Masonry

Ashlar coursed masonry consists of stones that are hammer dressed at the surface. The edges are not very perfectly squared. The stones are arranged in the courses. The depth of the courses lies between 15 cm to 30 cm. The depth of the stones are same in one course but can vary from curse to course.

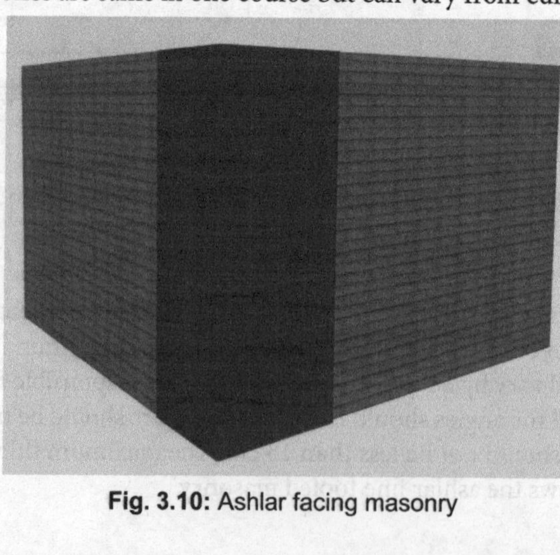

Fig. 3.10: Ashlar facing masonry

3.4.2.6 *Ashlar Facing Masonry*

Ashlar-facing masonry uses stones as the cornerstones while the rest of the wall is constructed by brick masonry or concrete blocks.

The bed and the sides are dressed properly for the stones and the surface is roughly tooled. The ashlar-facing masonry walls are aesthetically pleasing. Due to the sufficient use of the bricks in the backing, these types of walls are cost-effective. Figure 3.10 gives detail about the ashlar-facing masonry.

3.5 JOINTS IN STONE MASONRY AND THEIR PURPOSE

The stones are joined in the courses in different ways. The connection between two stones is known as a joint. The stones can be joined in different ways shown in figure 3.11. The various types of joints are explained as follows:

(1) **Butt joint:** The stones are placed side by side to form a butt joint. The butt joints are the simplest form of the joints in stone masonry. The stones are abutted to each other in such types of joints. This type of joint is also known as a square joint.

(2) **Lapped Joint:** Sometimes there may be movement between the stones. In this case, a lapped joint is provided. It is also called a rebated joint and it should not be less than 70 mm.

(3) **Tongued and grooved joint:** The tongued and grooved joint is a kind of interlocking joint. One part from a stone is projected and another stone has a grove. These two stones fit each other with the help of this projection and groove. This type of joint prevents the sliding of the stones.

(4) **Plugged joint:** The two stones to be joined are cut in the dove-tail shape and the molten lead is poured in this cut. This type of joint is known as a plugged joint. The plug holes can be filled with cement mortar too.

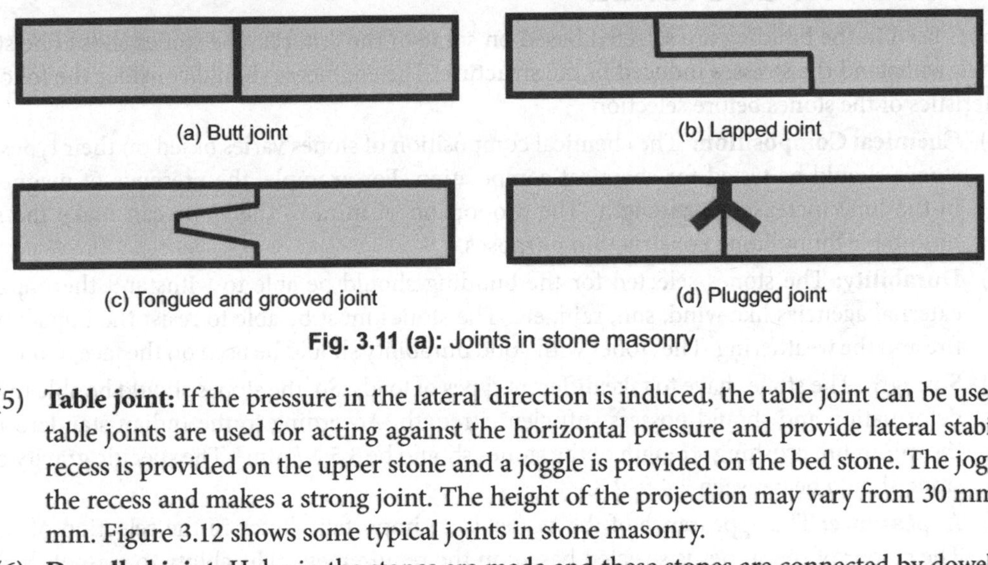

Fig. 3.11 (a): Joints in stone masonry

(5) **Table joint:** If the pressure in the lateral direction is induced, the table joint can be used. The table joints are used for acting against the horizontal pressure and provide lateral stability. A recess is provided on the upper stone and a joggle is provided on the bed stone. The joggle fits the recess and makes a strong joint. The height of the projection may vary from 30 mm to 40 mm. Figure 3.12 shows some typical joints in stone masonry.

(6) **Dowelled joint:** Holes in the stones are made and these stones are connected by dowels. The dowel is a piece of hard stone, slate, brass, etc. The joint is secured by cement. This type of joint is simple in nature. It provides stability against displacement and sliding.

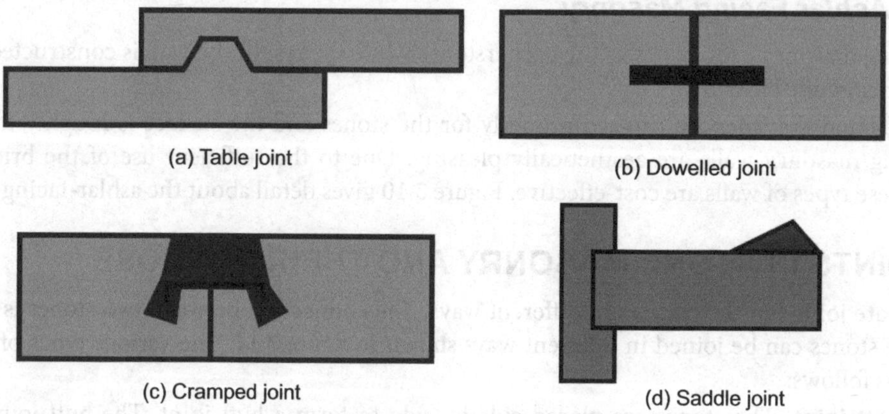

(a) Table joint (b) Dowelled joint (c) Cramped joint (d) Saddle joint

Fig. 3.11 (b): Typical joints in stone masonry

(7) **Cramped joints:** In a cramped joint, the holes in the adjacent stones are formed and the stones are connected by a metal cramp. Holes formed in the stones are dovetail shaped. The cramps can be made of steel, copper, or galvanized iron. The cramps must be free from corrosion. The ends of the cramp are turned down by 4 cm to 5 cm. The length of the cramp can be between 20 cm to 30 cm. The width can vary from 2 cm to 4 cm and the thickness can be kept between 5 mm to 10 mm. The holes can be grouted with cement or lead after placing the cramps.

(8) **Saddled joints:** The saddled joints are also known as the water joints. These are provided in the cornice generally and are used to divert the water on the slope.

3.6 SELECTION OF STONES

The stones used in the building are selected based on some of the criteria. The stones should be strong enough to withstand the stresses induced in the structure. The engineers should consider the following characteristics of the stones before selection:

(1) **Chemical Composition:** The chemical composition of stones varies based on their types. The stones should be tested for chemical composition. For example, the presence of magnesium in the lime increases its strength. The proportion of mica in the stone can make the stone unsuitable for building construction purposes.

(2) **Durability:** The stones selected for the building should be able to withstand the impact of external agencies like wind, sun, rain, etc. The stones must be able to resist the impact of the fire and the weathering. The stones with good durability should be used on the face of the walls.

(3) **Strength:** The stones have to take different types of loads. So, the stones should be able to resist deformation and should possess sufficient strength. According to the Indian standard code, the minimum crushing strength of the stones should be 3.5 N/ mm^2. The specific gravity of the stone should be between 2.4 to 2.8.

(4) **Appearance:** The appearance of the stones is an important factor in the selection of stones. The colour of the stones is selected based on the requirement. The ability to get polished and dressed affects the appearance of the stone. The stones with poor appearance can be used in the foundation works while the stones with pleasing appearance can be used in the surfacing works. Fine-grained stones with uniform texture have a better appearance. The grains should be compact and the stones should have a pleasing colour.

(5) **Hardness:** Hardness may be defined as the resistance to the localized deformation. The stones should be sufficiently hard. The coefficient of hardness should be at least 17 for pavements and 14 for building works.

(6) **Water Absorption:** The stones must not be porous in nature. There should not be excess pores in the stones. If the stone is immersed in the water for 24 hours, it should not absorb more than 5% water by its weight.

(7) **Cost:** The cost of the stones is another important factor. The dressing of the stones increases their cost. So, the stones are selected according to the budget too. The transportation charges of the stones are also very high.

3.7 PRECAUTIONS IN THE CONSTRUCTION OF STONE MASONRY

Stone masonry construction requires careful constriction due to the heavy stones and varying characteristics of the masonry. The following precautions should be taken during the construction of the stone masonry:

(1) The structure of the stones used in the construction must possess similar characteristics. These should be homogenous in nature.
(2) The stones should be sufficiently hard. Hardness is the property of the material to resist plastic deformation.
(3) The stones should be able to resist the shocks. The stones must be sufficiently tough in nature. The stones must possess sufficient strength to resist the loads.
(4) The dressing of the stones should be done according to the requirements.
(5) The stones should be watered properly. The moisture from the mortar should not be absorbed instantly by the stones.
(6) The stones should be checked for verticality using the plumb bob.
(7) Stones should be free from the defects like cavities and patches.
(8) The texture of the stones should be uniform in nature.
(9) The laying of stones should be at their natural bed. In this case, these will be more stable.
(10) The workability of the mortar must be checked properly.
(11) The continuity of the joints is avoided during the construction of the stone masonry.
(12) A proper bond should be maintained throughout the work during the stone masonry construction.
(13) The tensile stresses must not be induced as the stones possess low resistance towards the tensile forces.
(14) The cavities and depressions in the masonry should be filled with mortar.
(15) The minimum time required for the curing of the stone masonry is ten days.

3.8 BRICK MASONRY AND THE TERMS USED IN BRICK MASONRY

The stones are a natural material that may not be available everywhere. The quarrying and dressing of the stones is a complex procedure. So, brick masonry can prove a good alternative to stone masonry. In brick masonry, bricks are placed with mortar in a systematic manner to form a solid structure that can withstand the load. The size of modular brick is 20 × 10 × 10 cm with a layer of mortar. A civil engineer must know the various technical terms related to brick masonry. The different terms related to brick masonry are discussed as follows:

(a) **Header:** A Header may be defined as a brick that lies with its greater length perpendicular to the face of the work. It means the greatest length of the brick lies right angle to the work. In a modular brick, its face will measure 10 × 10 cm.

(b) **Stretcher**: A stretcher is a brick which laid in a way that its greater length along the face of the wall. Its means that greater length is parallel to the work. In a modular brick, its face will measure 20 × 10 cm. Figure 3.12 shows the header and stretcher.

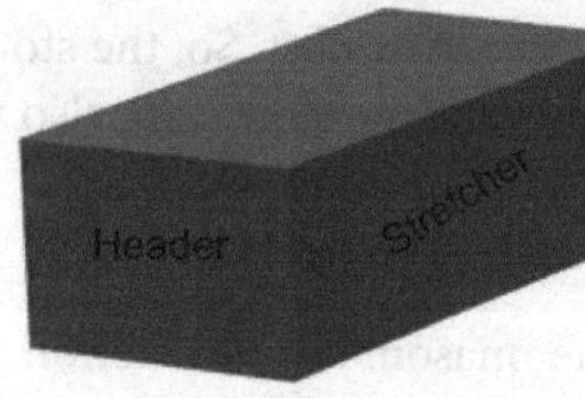

Fig. 3.12: Modular brick

(c) **Closer**: Closer is part of the brick in which one of the face of the greatest length remain uncut. The closer are obtained by breaking the bricks in a particular shape. The closers can be of different types depending on the shape.

(d) **King closer**: It is the part of the brick which is obtained by cutting the greater length and width by half the original size i.e. between the centers of length on one side to the center of width of one side. In this triangular piece is cut from a brick. Figure 3.13 (a) shows the king closer.

(e) **Queen closer:** It is the part of the brick which is obtained by cutting the brick into two equal half along the lengthwise. Queen closer brick is half the width of the modular brick. Figure 3.13 (b) shows the queen closer.

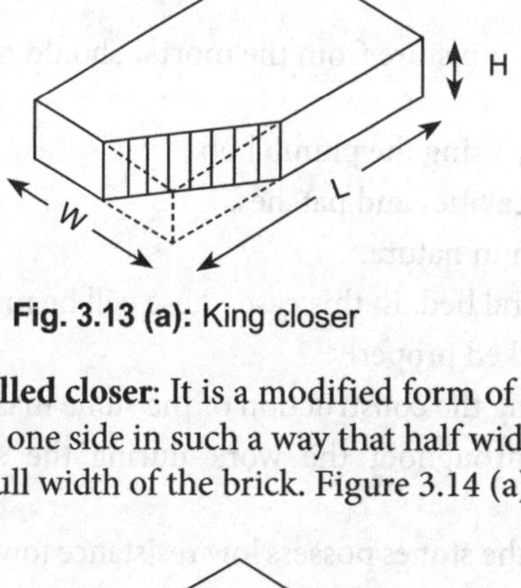

Fig. 3.13 (a): King closer

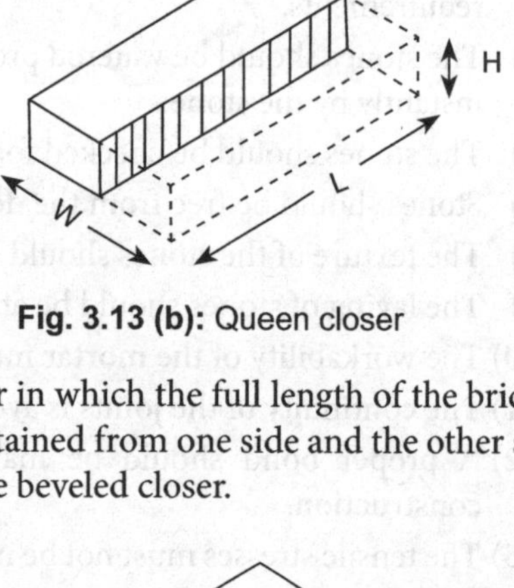

Fig. 3.13 (b): Queen closer

(f) **Bevelled closer**: It is a modified form of king closer in which the full length of the brick is cut from one side in such a way that half width is maintained from one side and the other side has the full width of the brick. Figure 3.14 (a) shows the beveled closer.

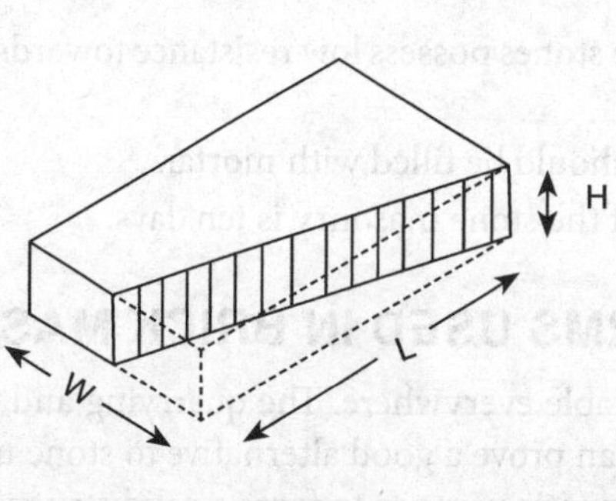

Fig. 3.14 (a): Bevelled closer

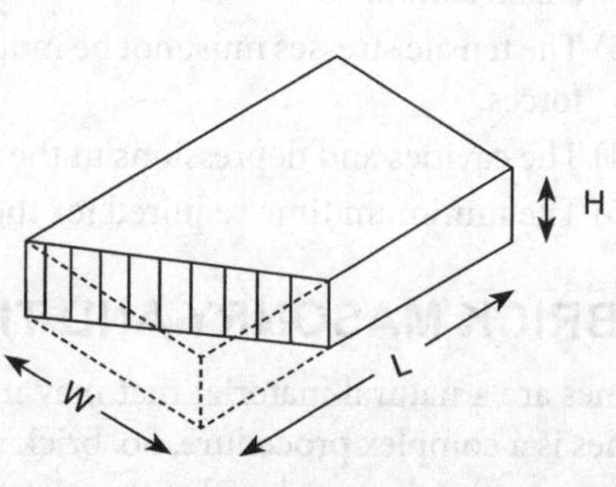

Fig. 3.14 (b): Mitred closer

(g) **Mitred closer**: It is a brick that can be obtained by splaying from one end at an angle between 45° to 60°. In these brick one side of the length is equal to the original size while the other half is cut from the center of the length. Figure 3.14 (b) shows the mitred closer.

(h) **Quoins**: Brick masonry that is present at the corner of the wall is known as quoins. They provide an aesthetic look as well as strength. They also provide protection from the weather in a load-bearing structure. Quoins are of header and stretcher type.

(i) **Quoin Header:** When the brick of quoin is laid in such a way that its greater length is perpendicular to the face of the wall, it is called quoin header.

(j) **Quoin Stretcher:** When the brick of quoin is laid in such a way that its greater length is parallel to the face of the wall.

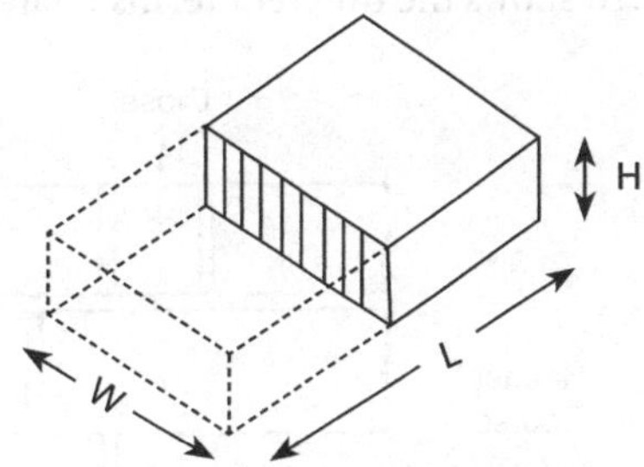

Fig. 3.15 (a): Half bat

(k) **Course:** A Course can be defined as a horizontal layer running along the wall. In brick masonry, course thickness is equal to the height of modular brick with one mortar joint. The course can be of header and stretcher type. In a header course, all the bricks will show the header as the face of the wall.

(l) **Bat:** It is part of the brick that is cut perpendicular to the length of the brick. Therefore, the length of the bat brick is smaller than modular brick. The bats can be half bat, three-quarter bat or beveled bat. In a half-bat, the length of the brick is half of the length of the modular brick. In a three-quarter bat, the length of the brick is equal to the three-quarter to length of the modular brick. In the beveled bat, the brick is beveled. Figure 3.15 shows the details of different types of bat.

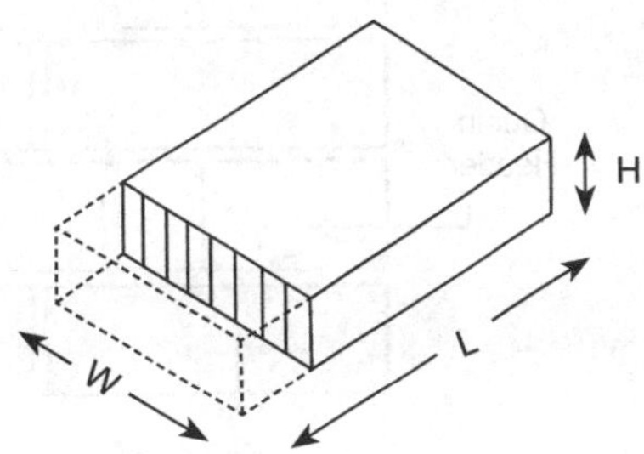

Fig. 3.15 (b): Three-quarter bat

(m) **Bond:** It is an art by which bricks stick to each other. It helps in arranging bricks in the course by connecting them together. Bond makes the wall look aesthetically pleasing.

(n) **Joints:** A joint is defined as the junction between the adjacent bricks. When joints are present in parallel along the face of the wall is known as a bed joint. If the joints are present at the right angle to the wall, these are known as vertical joints.

(o) **Lap:** A lap is defined as the distance along the face of the all between two vertical joints in consecutive courses. The minimum length of the lap for a good brick joint is 1/4th of the length of a modular brick.

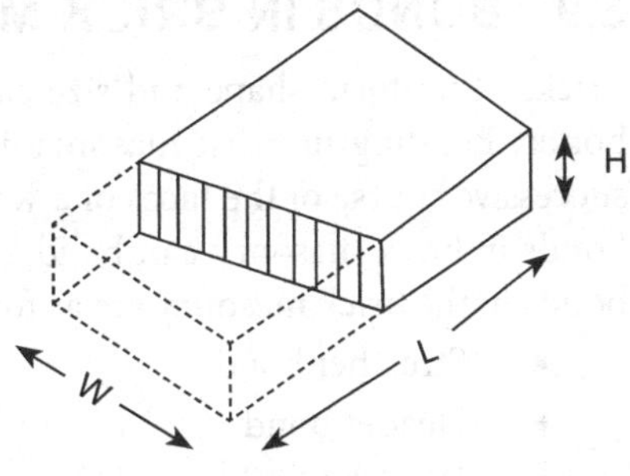

Fig. 3.15 (c): Bevelled bat

(p) **Frog Line:** A frog line is a depression on one side of the face of the brick. It is provided in the brick to form a joint with the mortar and prevent the above brick from displacement.

(q) **Level:** level is used to keep the courses of brick masonry horizontal.

(r) **Plumb:** Plumb is used for keeping the face of the wall vertical.

The terms like hearting, facing and backing are explained in terms used in stone masonry. Figure 3.16 shows the different terms related to the brick masonry implemented in a wall.

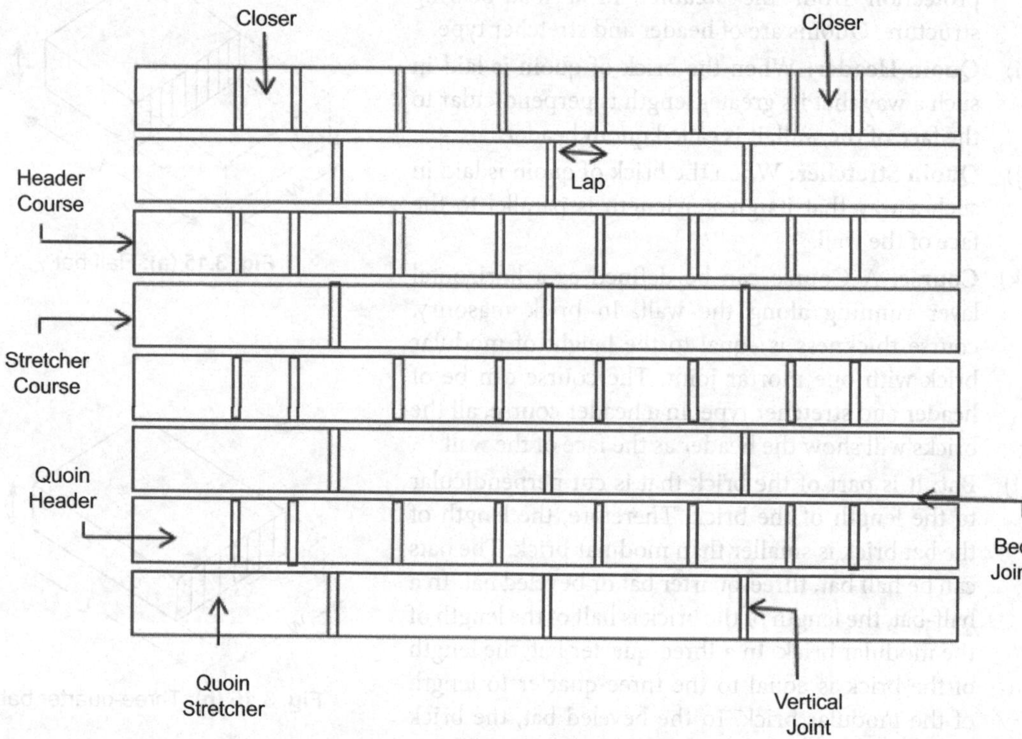

Fig. 3.16: Terms used in brick masonry

3.9 BONDS IN BRICK MASONRY

Bricks of uniform shape and size can be arranged in different patterns that result in various types of bonds. Bonding in brick masonry is necessary to remove the continuity of the vertical joints in the successive course of the faces of a wall. The bonding provides strength to the wall. The various types of bonds in brick masonry can be identified by the plan and elevation of the wall. The different types of bonds in the brick masonry are as follows:

- Stretcher bond
- Header bond
- English bond
- Flemish bond

3.9.1 STRETCHER BOND

This type of bond is made by laying all bricks in a horizontal course such that all the stretchers appear on the wall facing. The length of the brick is parallel to the horizontal direction. A stretcher bond can only be used when a wall has the thickness of half brick. The size of standard brick is 19cm × 9 cm× 9cm. It means that the wall having a thickness of 9 cm can only be constructed using a stretcher bond. This pattern is generally used for the construction of partition walls, curtain walls, and chimney stacks. The different views of the stretcher bond are shown in figure 3.17.

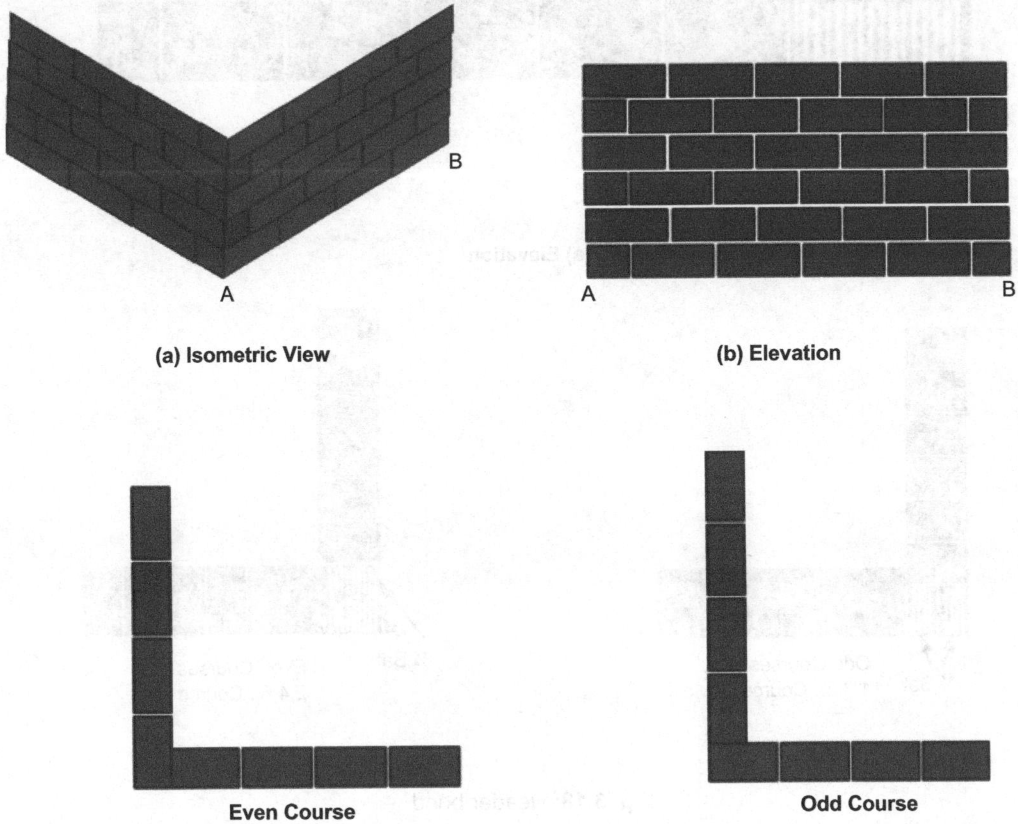

Fig. 3.17: Terms used in brick masonry

3.9.2 Header Bond

This type of pattern can be obtained by placing all bricks as header on the wall facing. The length of brick is perpendicular to the horizontal direction. Header bond is used when the thickness of wall is equal to full brick thickness i.e. 18 cm. Half brick thickness i.e. 9 cm is kept as an overlap. It can be managed by using three quarter bat as quoins in alternatives course. This type of pattern cannot transfer the pressure or load of building along the length of wall. This bond is not used for load bearing wall. Header bond is specially used where design required curved surfaces as stretcher bond cannot be used because stretcher bond will project outside the face of the wall. Header bond can also use in foundation for brickwork. Figure 3.18 shows different views of the header bond.

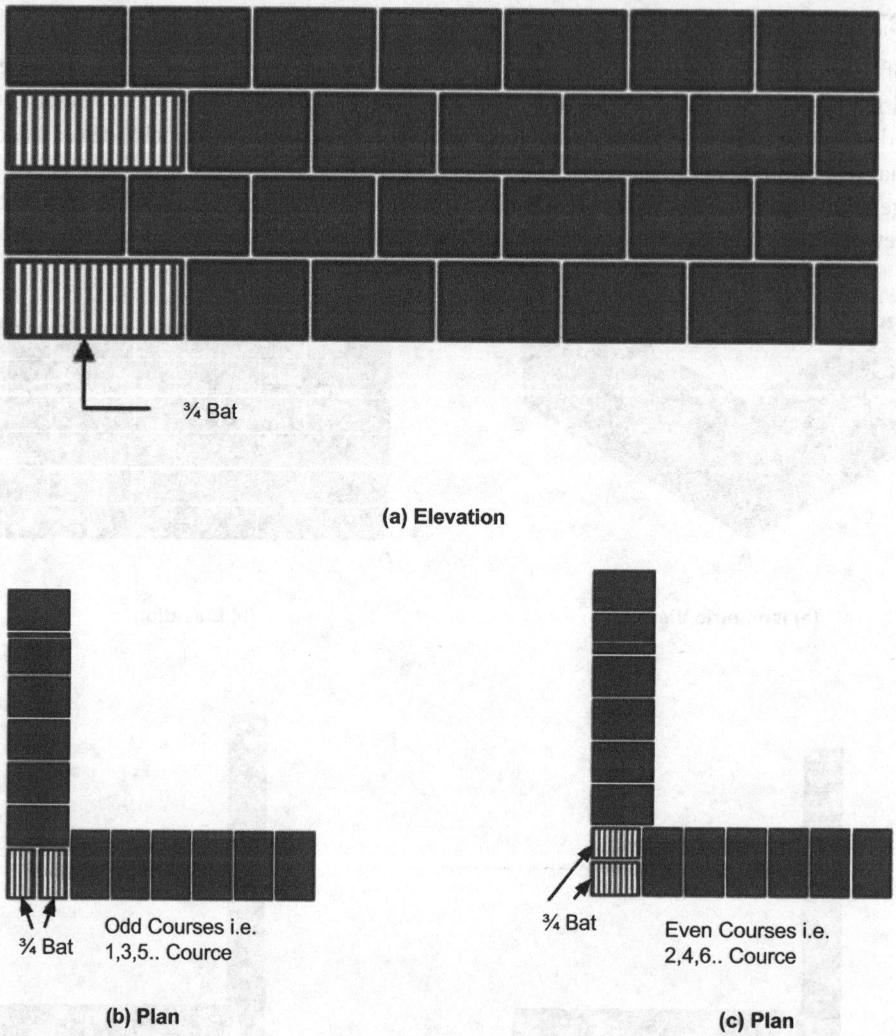

Fig. 3.18: Header bond

3.9.3 English Bond

Header and stretcher course are present alternatively in English bond. In this type of bond, the vertical joints present in the header course and stretcher course come in the same line throughout the wall. For disturbing the continuity of vertical joints in the consecutive course it is necessary to have the queen closer after the header quoin in each header course. Half bat or three-quarter bat brick is not used in the construction of English bond. This bond is simple to construct and stronger than all other bonds. Wall can be constructed fast by using this bond as compared to other bonds. Walls of any thickness can be constructed using English bond. Different arrangements of English bond for even and odd course for different thickness is shown in figure 3.19. Queen closer is shown in the shaded form. Queen closers help in breaking the continuity of vertical joints.

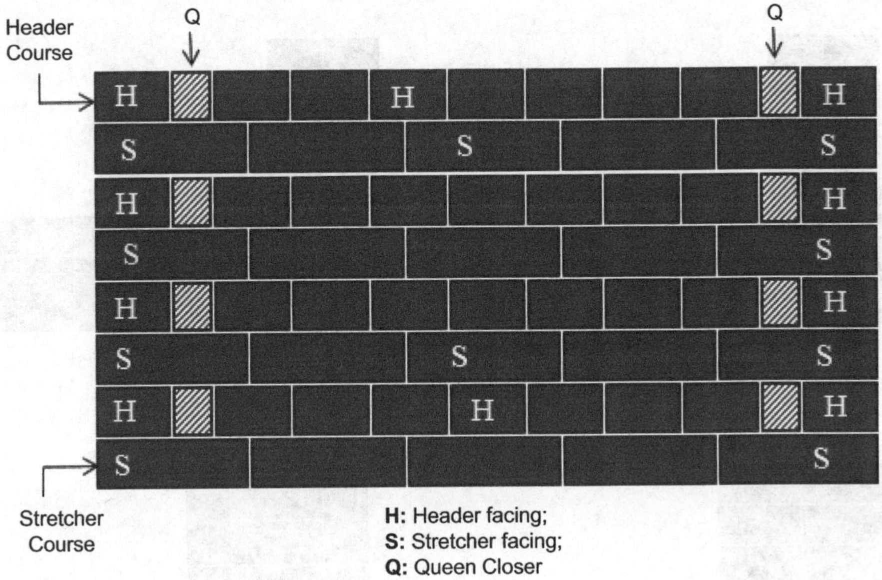

Fig. 3.19: English bond

Some additional points for the construction of English bond masonry are given as follows:

(1) A queen closer is never used at the start or at the end of the header course as it will easily lodge from its position.

(2) In stretcher courses, the minimum overlap of the stretcher is kept 1/4th the length of the brick over the header.

(3) A wall with a thickness equal to an even number of half bricks (1 brick, 2 brick, 3 brick thick wall, and so on) will have a similar appearance on the outer and inner faces of the wall i.e. a header course will have header face on the front and back face. A similar situation will occur in the stretcher course. Figures 3.20 (a), (b) and (c) show the plan of different brick thickness.

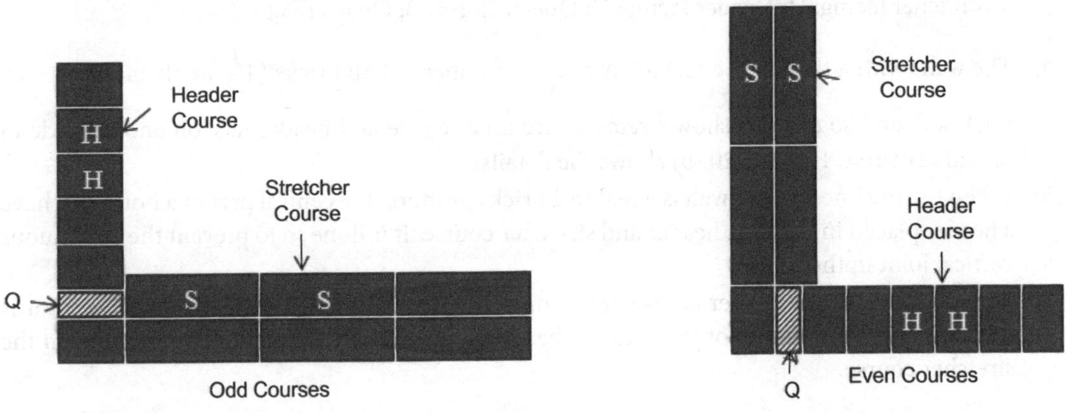

Fig. 3.20 (a): Plan for one brick thick wall

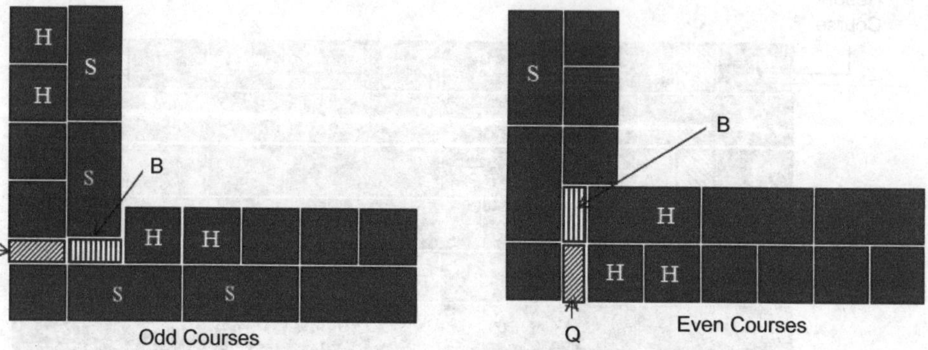

Fig. 3.20 (b): Plan for one and half brick thick wall

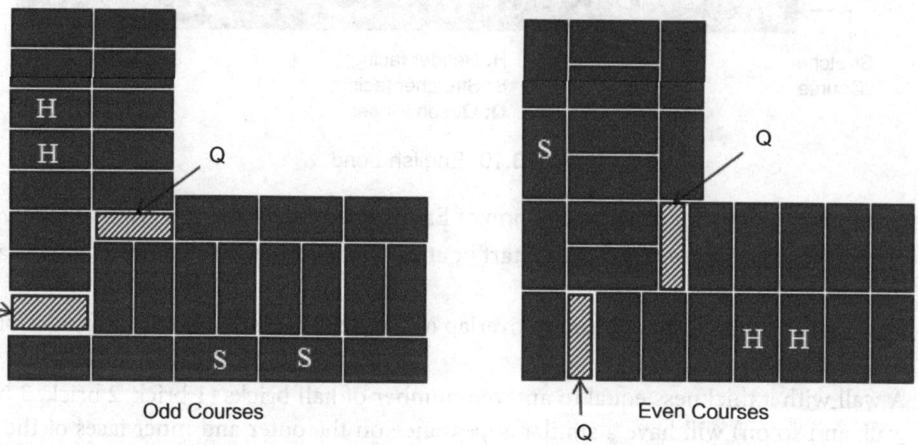

Fig. 3.20 (c): Plan for two brick thick wall

S; Stretcher facing; H: Header facing; Q: Queen closer; B: Quarter bat

(4) The walls with a thickness equal to an uneven number of half bricks ($1\frac{1}{2}$ brick and $2\frac{1}{2}$ brick thick wall and so on) will show stretcher face on one side and header face on another side in the same course. Figure 3.20 (b) shows the details.

(5) When the thickness of the wall is equal to 2 bricks or more, the central part of a bond will have a header placed in both the header and stretcher course. It is done so to prevent the continuous vertical joint in the wall

(6) Vertical joints in the header course are thinner than vertical joints in the stretcher joint it is done because the number of joints in the header course is double the number of joints in the stretcher course.

3.9.4 Flemish Bond

In Flemish bond, the header and stretcher are present alternatively in the same course, and the appearance remains the same in every course of the brick masonry. In this arrangement, the alternative courses start with a header and stretcher face. Queen closer is present after header quoin in the course which starts with header face for disturbing the vertical joints continuity in the successive courses. Bat is used in this type of bond when the thickness of the wall is equal to the odd multiple of half brick ($1\frac{1}{2}$ brick and $2\frac{1}{2}$ brick thick wall and so on). The Flemish bond wall is shown in figure 3.21. The Flemish bond appears more aesthetically pleasing than the English bond but the English bond is stronger than the Flemish bond.

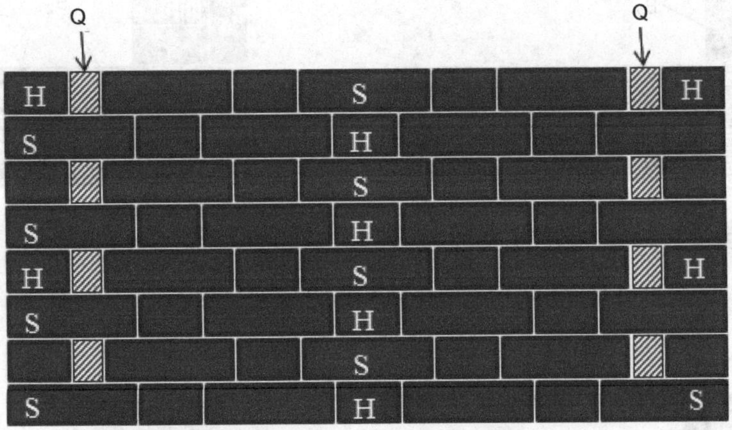

Fig. 3.21: Elevation of Flemish bond

The Flemish bonds are further classified into two types:

- **Double Flemish bond:** In this arrangement of brick masonry, every course has a Flemish bond on the front and back elevation of the wall. Half bat bricks are used in $1\frac{1}{2}$ brick and $2\frac{1}{2}$ brick thick wall construction which makes them weaker than English bond. The header and stretcher are placed alternatively in each course. Double Flemish bonds for different thicknesses and for even and odd courses are shown in figure 3.22 (a) to (c).

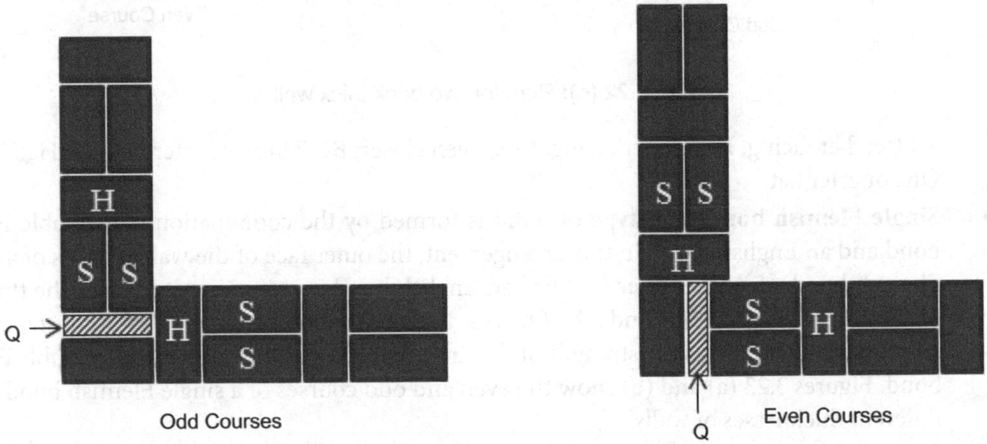

Fig. 3.22 (a): Plan for one brick thick wall

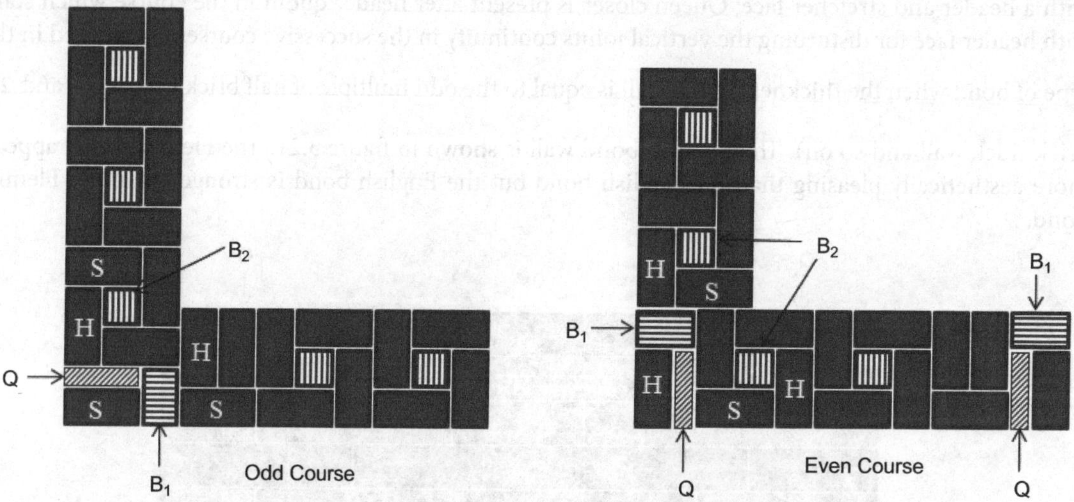

Fig. 3.22 (b): Plan for one and half brick thick wall

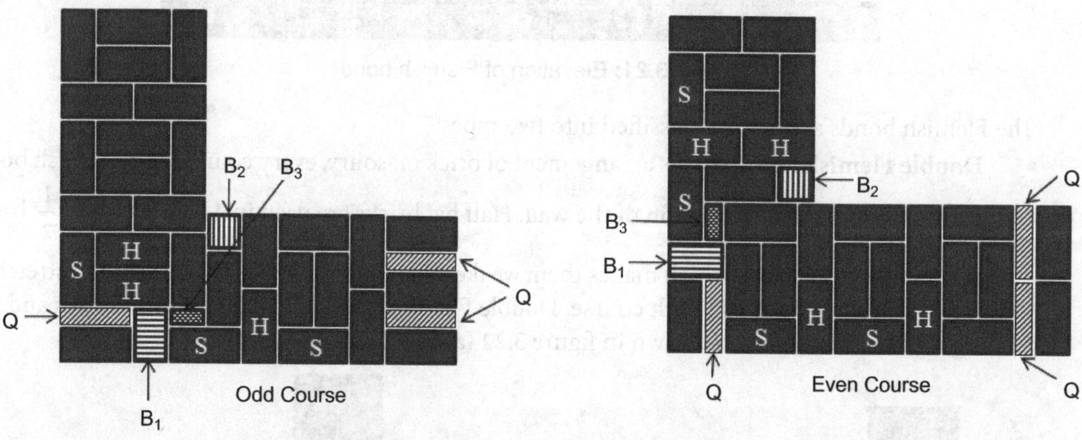

Fig. 3.22 (c): Plan for two brick thick wall

S: Stretcher facing; H: Header facing; Q: Queen closer; B1: Three-quarter bat; B2: Half bat; B3: One quarter bat

- **Single Flemish bond**: This type of bond is formed by the combination of a double Flemish bond and an English bond. In this arrangement, the outer face of the wall consists of a double Flemish bond while the inner face has an English bond in every course. When the thickness of the wall is less than one and a half bricks, this bond cannot be adopted for construction. A single Flemish bond has the strength of the English bond and the beauty of the double Flemish bond. Figures 3.23 (a) and (b) show the even and odd courses of a single Flemish bond for the different thicknesses of walls.

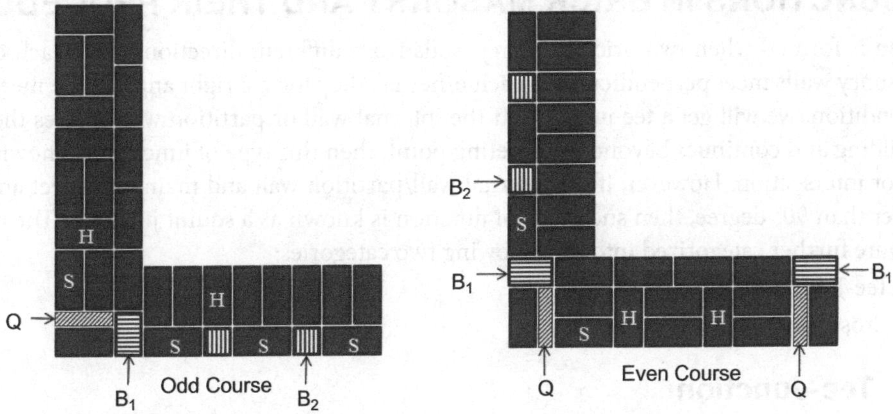

Fig. 3.23 (a): Plan for one and half brick thick wall

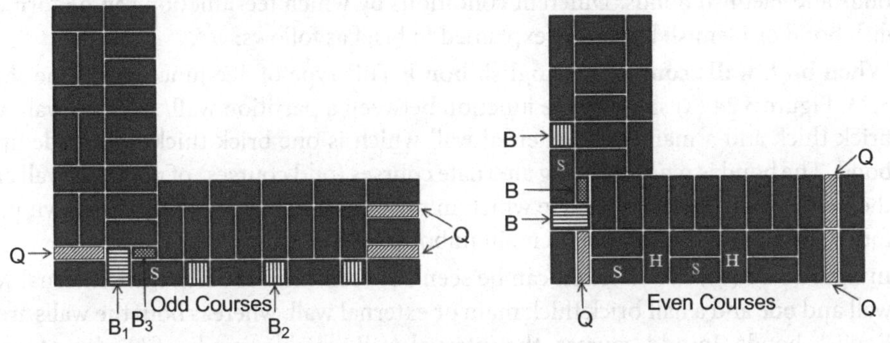

Fig. 3.23 (b): Plan for two brick thick wall

S: Stretcher facing; H: Header facing; Q: Queen closer; B1: Three-quarter bat; B2: Half bat; B3: One quarter bat Brick

3.10 REQUIREMENTS OF A GOOD BRICK

A good brick has the following requirements:

- Bricks used in masonry should be uniform in size and shape. No cracks or fractures present on the surface of the brick
- All brick courses should be horizontal and all joints should be vertical.
- Brick bats should only be used where they are required
- Bricks should be laid in such a way that the frog is always present on top.
- Brick masonry should be such that it can resist the lateral or sideway thrust
- It should be strong enough to distribute the loads throughout the wall
- The appearance of the brocks in the walls should be aesthetically pleasing.
- Brick masonry should be durable and have a long life
- The bricks should be fire-resistant.
- The bricks should act as a barrier to sound.

3.11 JUNCTIONS IN BRICK MASONRY AND THEIR PROCEDURE

A Junction is formed when two brick masonry walls from different directions meet each other. If the brick masonry walls meet perpendicular to each other i.e. they form a right angle at the meeting points. In this condition, we will get a tee junction. If the internal wall or partition wall crosses the main wall of the building and continues beyond the meeting point, then this type of junction is known as a cross-junction or intersection. However, if the internal wall/partition wall and main wall meet and form any angle other than 90° degree, then such type of junction is known as a squint junction. The right angled junctions are further categorized into the following two categories:

- Tee-Junction
- Cross-junction or intersection

3.11.1 Tee-Junction

Tee-junction in brick masonry is formed when the partition wall/ internal wall meets the main wall/ external wall at a 90° angle. Tee-junction can be formed by various thicknesses of walls made up of English bonds and Flemish bonds. Different conditions by which tee-junction can be formed by either using English bond or Flemish bonds are explained in brief as follows:

(1) **When both walls contain an English bond:** This type of Tee junction can be seen in figure 3.24. Figure 3.24 (a) shows a Tee junction between a partition wall/ internal wall which is half brick thick and a main wall/ external wall which is one brick thick wall made up of English bond. The bond is made by using alternate courses (odd courses) of partition wall entering into the main wall of stretcher course which in turn helps in overlap which is shown in the shaded form. Even courses of the wall remain unbounded.

In figure 3.24 (b), Tee-junction can be seen between the one brick thick internal wall or cross wall and one and a half brick thick main or external wall, whereas both the walls are built using English bonds. In odd courses, the internal wall which is made of header courses is placed upon the stretcher courses of the external wall through half of the brick width shown in the shaded form. Queen closer (Q) is used for overlap by which tie brick (shaded) is overlapped by quarter brick width. The even courses of both the walls remain unbounded.

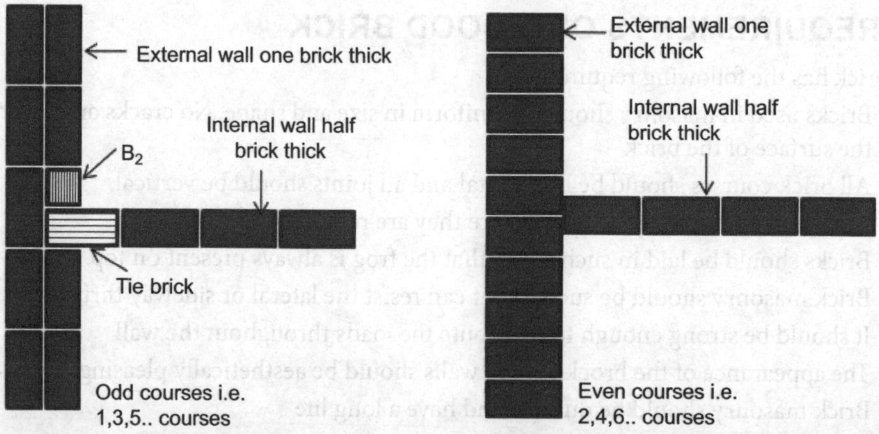

Fig. 3.24 (a): Tee-junction between half brick thick wall and one brick thick wall

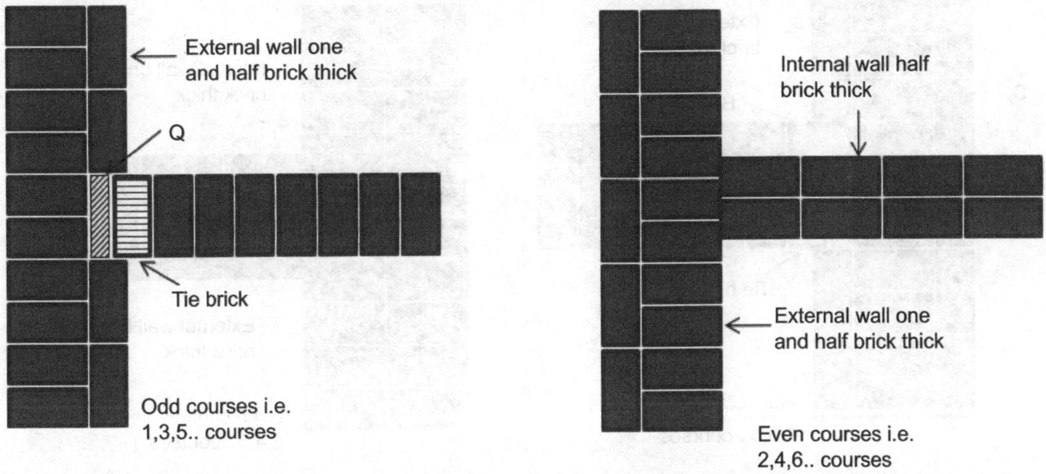

Fig. 3.24 (b): Tee-junction between one and a half brick thick wall and one brick thick wall

B2; Half bat brick; B1: Three-quarter bat brick; Q; Queen closer

In figure 3.24 (c), Tee-junction can be seen between the one and a half brick thick internal wall or cross wall and one and a half brick thick main or external wall, whereas both walls are built using English bond. In odd courses, the header brick of the internal wall enters into the junction with stretcher course of the external wall. The overlap of a tie brick (shaded brick) is obtained by using queen closer (Q). Three- fourth quarter brick (B1) is used for additional overlap in the same course. Even courses of both wall remains unbounded.

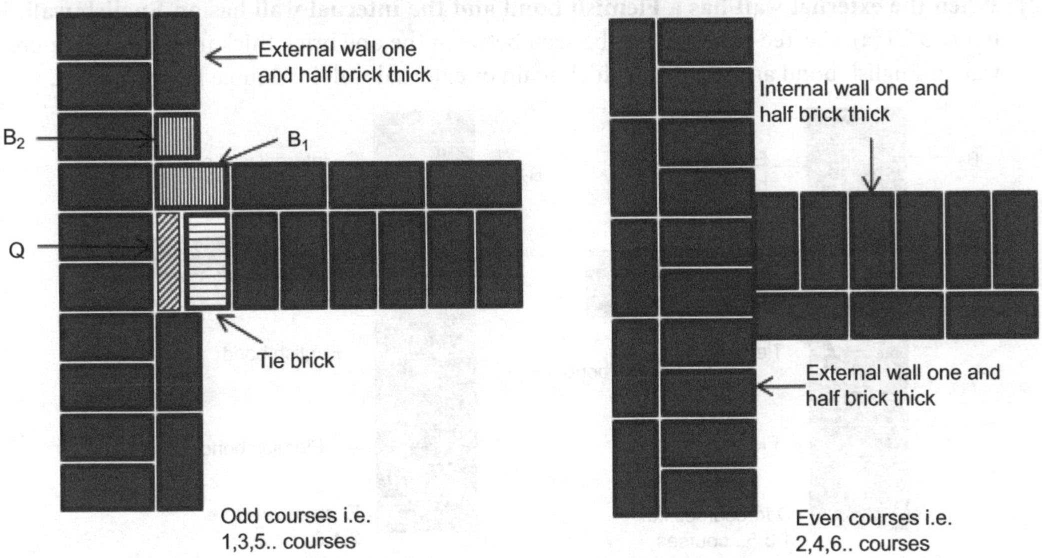

Fig. 3.24 (c): Tee-junction between two one and a half brick thick walls

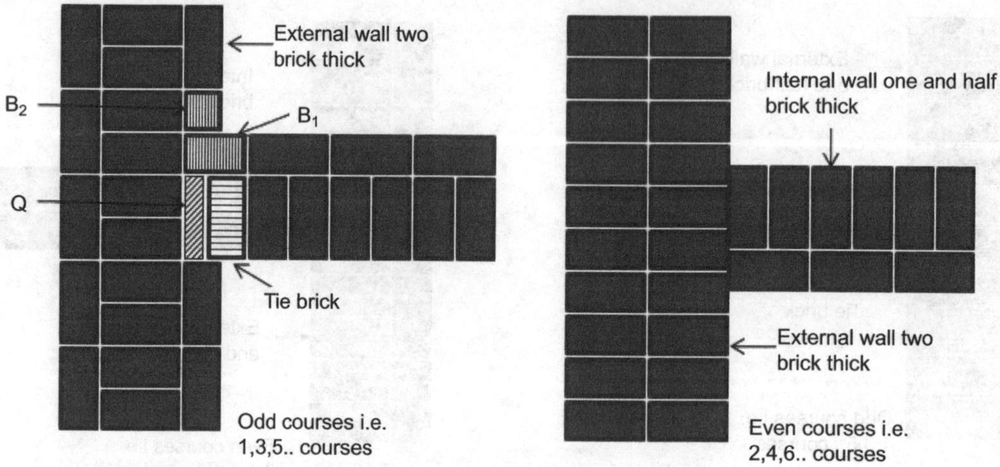

Fig. 3.24 (d): Tee-junction between one and a half brick thick wall and two brick thick wall

B2; Half bat brick; B1: Three-quarter bat brick; Q; Queen closer

In figure 3.24 (d), Tee-junction can be seen between the one and a half brick thick internal wall or cross wall and two brick thick main or external wall, whereas both walls are built using an English bond. In odd courses, the header course of the internal wall enters into the junction with the stretcher course of the external wall through half of its width. Queen closer (Q) is used for the overlap by which tie brick (shaded) which is placed as header brick is overlapped by a quarter-brick width. Three- fourth quarter brick (B1) is used for additional overlap in the same course. Here also, even courses of both wall remains unbounded.

(2) **When the external wall has a Flemish bond and the internal wall has an English wall:** In figure 3.25 (a), the Tee-junction can be seen between the one brick thick internal wall or cross wall in English bond and one brick thick main or external wall in Double Flemish bond.

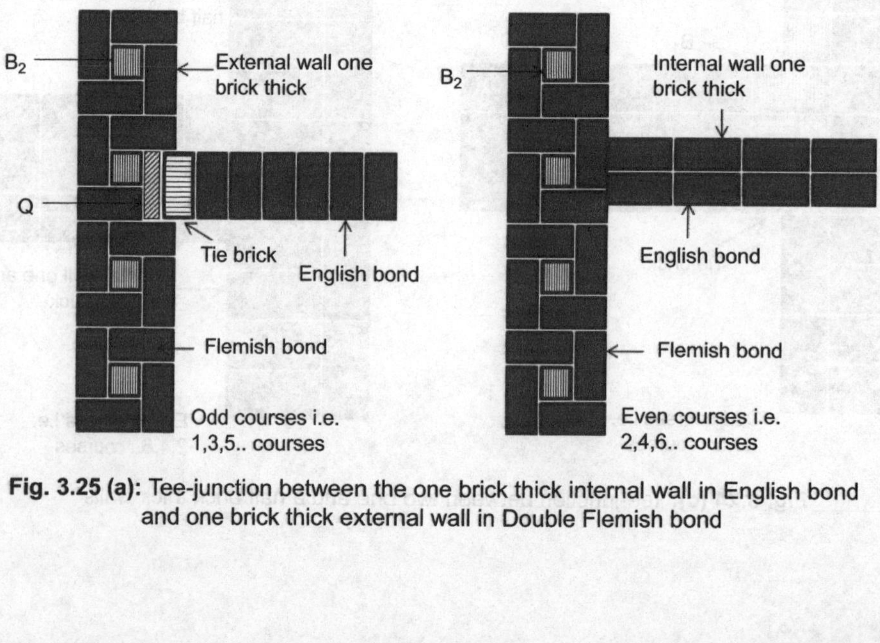

Fig. 3.25 (a): Tee-junction between the one brick thick internal wall in English bond and one brick thick external wall in Double Flemish bond

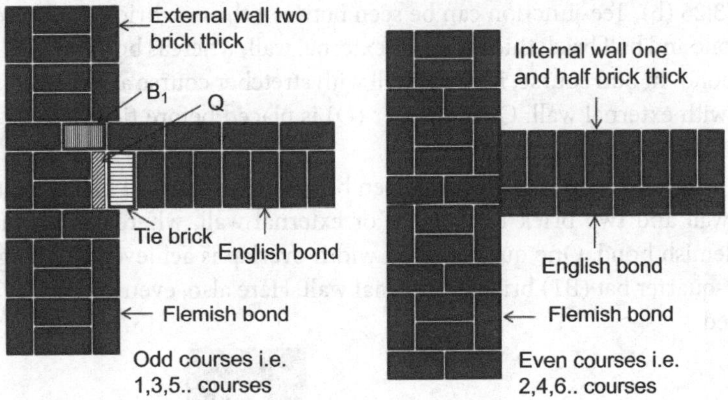

Fig. 3.25 (b): Tee-junction between the one and a half brick thick internal wall in English bond and Two brick thick external wall in Double Flemish bond

B2; Half bat brick; B1: Three-quarter bat brick; Q; Queen closer

In the odd course, the internal wall with the header course enters into the junction with the external wall causing a quarter-brick width overlap. Queen closer (Q) is placed before tie brick (shaded) of header course in an internal wall. The even courses of both the walls remain unbounded.

In figure 3.25 (b), Tee-junction can be seen between the one and a half brick thick internal wall or cross wall in English bond and Two brick thick main or external wall in Double Flemish bond. Here also, the internal wall with the header course enters into the junction with the external wall causing a quarter-brick width overlap. Queen closer (Q) is placed before the tie brick (shown in shaded form) of the header course in the internal wall. Three-quarter bat brick (B1) is placed next to the stretcher brick of the main wall for an additional overlap which is in the same course as the internal wall. The even courses of both the walls remain unbounded.

(3) **When both the walls are constructed by double Flemish bond:** In figure 3.26 (a), Tee-junction can be seen between the half brick thick internal wall or cross wall and one brick thick main or external wall, whereas both the wall built using double Flemish bond. In odd course, internal wall with stretcher courses enter into the junction with external wall through half brick length. Half bat brick (B2) is placed in the external wall due to this overlapping. Even courses of both wall remains unbounded.

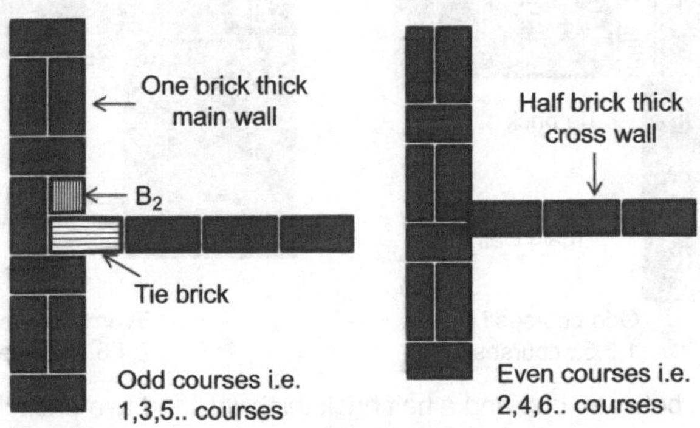

Fig. 3.26 (a): Tee junction between one brick thick wall and half brick thick wall in Flemish bond

In figure 3.26 (b), Tee-junction can be seen between the one brick thick internal wall or cross wall and one and half brick thick main or external wall, whereas both the wall built using double Flemish bond. In odd course, internal wall with stretcher course as tie brick (shaded) enter into junction with external wall. Queen closer (Q) is placed before tie brick for overlapping. Even courses of both wall remains unbounded.

In figure 3.26 (c), Tee-junction can be seen between the one and half brick thick internal wall or cross wall and two brick thick main or external wall, whereas both the wall built using double Flemish bond. One quarter brick width overlap is achieved by using queen closer (Q) and three-quarter bat (B1) brick in external wall. Here also, even courses of both wall remains unbounded.

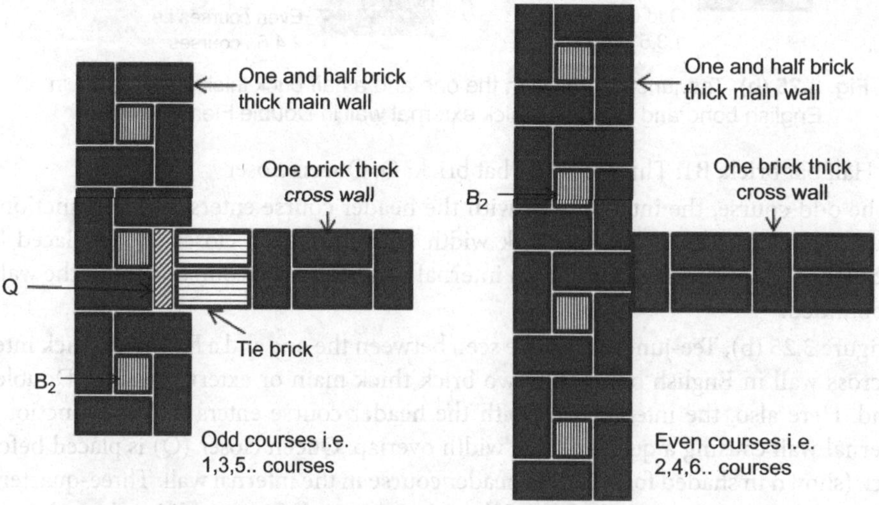

Fig. 3.26 (b): Tee junction between one and a half brick thick wall and one brick thick wall in Flemish bond

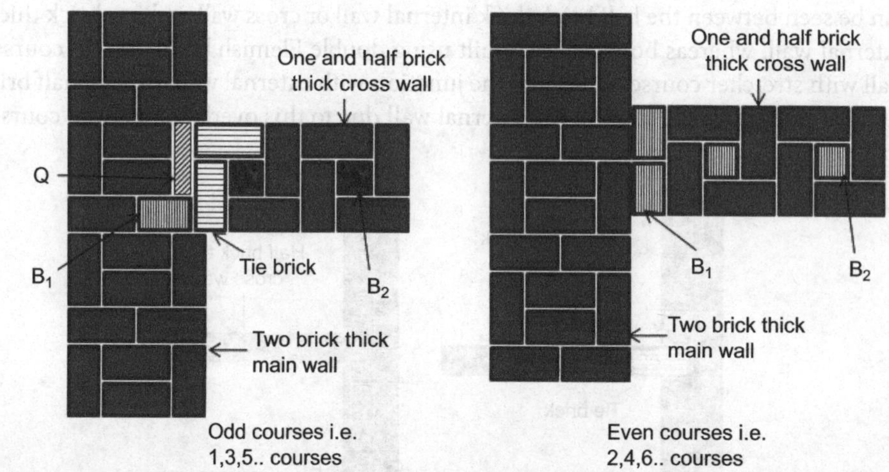

Fig. 3.26 (c): Tee junction between one and a half brick thick wall and two brick thick wall in Flemish bond

B2; Half bat brick; B1: Three-quarter bat brick; Q; Queen closer

3.11.2 Intersection or cross-junction

An intersection or cross-junction is created when two walls are crossed perpendicular to each other i.e. they formed right angle at junction. In this junction one wall will called as main wall while the other wall with less thickness is known as cross wall. In figure 3.27 (a), a cross-junction can be seen between the one brick thick internal wall or cross wall and one and half brick thick main wall, whereas both the wall built using English bond. In odd course, cross-wall having header course enter into junction with main wall. Tie bricks are used in cross-wall to achieve an overlap of quarter brick width on both sides. Even courses of both wall remains unbounded.

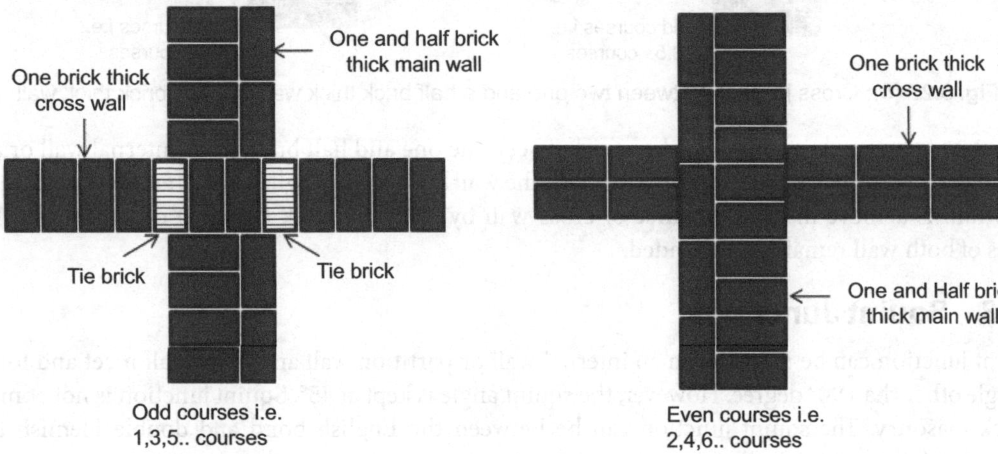

Fig. 3.27 (a): Cross junction between one and a half brick thick wall and one brick thick wall

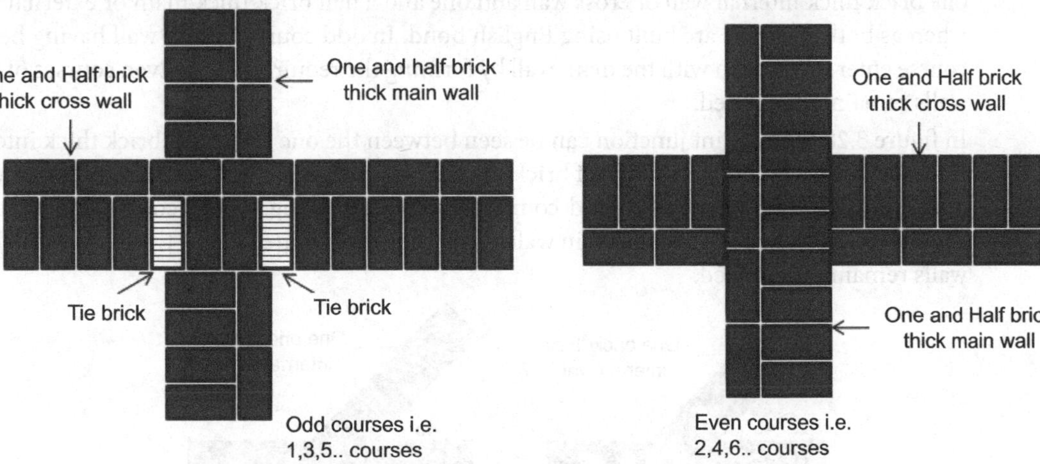

Fig. 3.27 (b): Cross junction between two one and a half brick thick walls

In 3.27 (b), a cross-junction can be seen between the one and half-brick thick both internal wall or cross wall and main wall, whereas both the walls built using English bond. In odd course, an overlap of quarter brick width is achieve in header course of cross-wall by using tie brick (shaded) on both faces. Even courses of both walls remain unbounded.

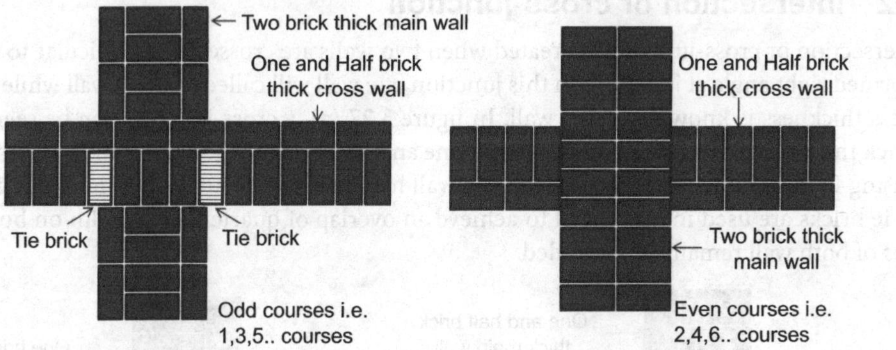

Fig. 3.27 (c): Cross junction between two one and a half brick thick wall and two brick thick wall

In 3.27 (c), a cross-junction can be seen between the one and half brick thick internal wall or cross wall and two brick thick main wall, whereas both the wall built using English bond. an overlap of quarter brick width is achieve in header course of cross-wall by using tie brick (shaded) on both faces. Even courses of both wall remains unbounded.

3.11.3 Squint Junctions

A squint junction can be made when an internal wall or partition wall and main wall meet and formed any angle other than 90° degree. However, the squint angle is kept at 45°. Squint junction is not common in brick masonry. The squint junction can be between the English bond and double Flemish bond explained as follows:

(1) **Squint junction in English Bond:** In figure 3.28 (a), a squint junction can be seen between the one brick thick internal wall or cross wall and one and a half brick thick main or external wall, whereas both the walls are built using English bond. In odd courses, cross-wall having header course enter in junction with the main wall by forming the required angle. Even courses of both walls remain unbounded.

In figure 3.28 (b) a squint junction can be seen between the one and a half brick thick internal wall or cross wall and one and a half brick thick main or external wall, whereas both the walls built using an English bond. In odd courses, cross-wall having header course and stretcher course enter in junction with the main wall by forming the required angle. Even courses of both walls remain unbounded.

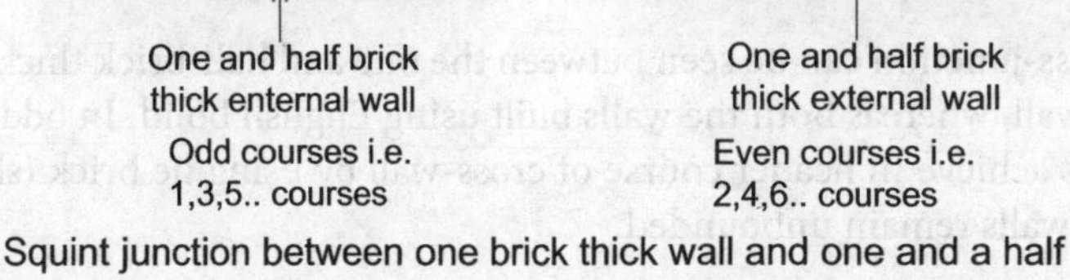

Fig. 3.28 (a): Squint junction between one brick thick wall and one and a half brick thick wall

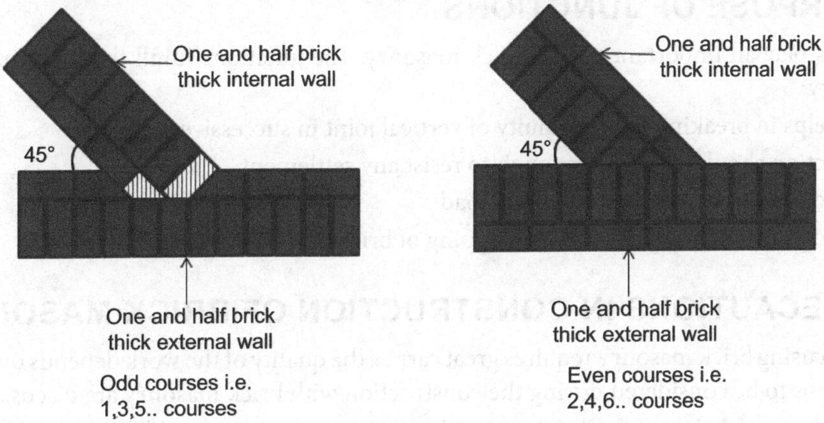

Fig. 3.28 (b): Squint junction between one and half brick thick walls

(2) **Squint junction in double Flemish Bond:** In figure 3.29 (a), a squint junction can be seen between the one brick and one and half brick thick internal wall or cross wall and one and half brick thick main or external wall, whereas both the wall built using double Flemish bond. It is difficult to construct squint junction in double Flemish bond. In figure 3.29 (b), a squint junction is shown between tow walls of one and a half brick thickness.

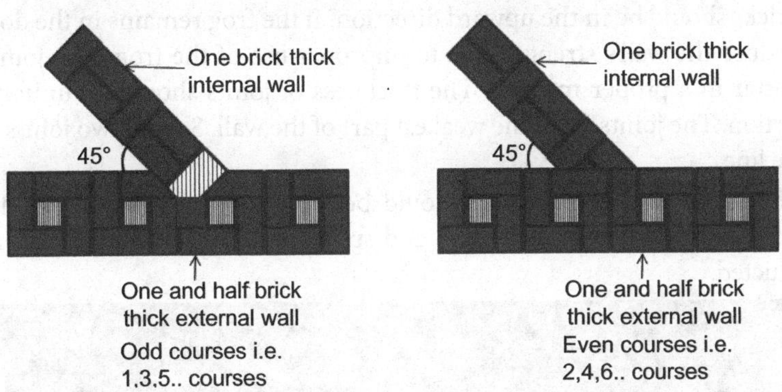

Fig. 3.29 (a): Squint junction between one brick thick wall and one and a half brick thick wall

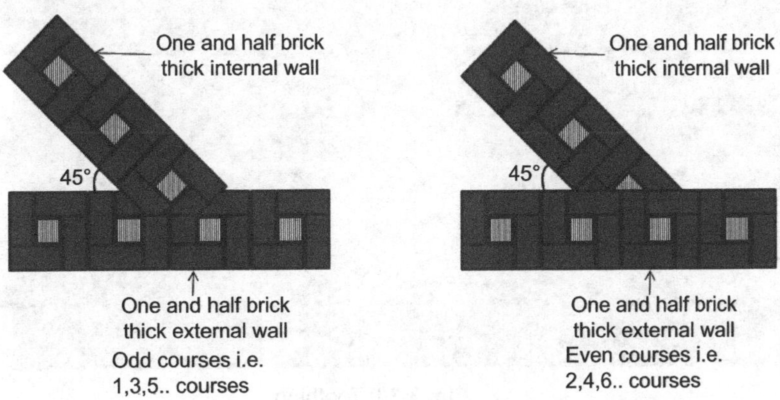

Fig. 3.29 (b): Squint junction between two one and a half brick wall

3.12 PURPOSE OF JUNCTIONS

The junctions play an important role in brick masonry. The junctions fulfill the following purposes in brick masonry:

- It helps in breaking the continuity of vertical joint in successive course
- Junction should be strong enough to resist any settlement.
- It helps in transferring the vertical load
- It helps in achieving desired overlapping of bricks

3.13 PRECAUTIONS IN CONSTRUCTION OF BRICK MASONRY

Construction using brick masonry requires great care as the quality of the work depends on many factors. The precautions to be considered during the construction with brick masonry are discussed as follows:

(1) **Soaking of bricks:** The bricks used in the construction should be saturated in the water properly. The bricks should be soaked in water at least for a duration of 12 hours. However, the soaking time depends on the quality of the bricks. If the bricks are not saturated, they can absorb water from the mortar which can make the bond weak.

(2) **Beds and joints:** The bottom surface of the bricks should lie on the mortar properly. There should not be any inclination of the bed with respect to the mortar line. The pressure coming to the brick must be perpendicular to its bed. The bricks should be laid on their bed and the frog of the bricks should be in the upward direction. If the frog remains in the downward direction, it will reduce the wall's strength due to improper use of the frog. The joints should be filled with mortar in a proper manner. The thickness of joints should be uniform throughout the construction. The joints form the weakest part of the wall. So, the two joints should never be in the same line.

(3) **Verticality:** The wall constructed should be truly vertical. The verticality of the wall can be tested with the help of plumb bob and spirit level. If the wall is not vertical, it should be reconstructed.

Fig. 3.30: Toothing

(4) **Wetting of bed:** If the construction of the wall begins from an already constructed part, the bed of the previously constructed part is to be wetted with water properly. The water pipe can be used for wetting the bed of previous construction. It will help in establishing a strong bond between new and old work.

(5) **Toothing:** If there are plans for construction in the future, the toothing can be provided. When the walls cross each other, the extension of bricks is provided in the wall so that when future work begins it can make a good bond with old construction. The figure 3.30 shows the toothing.

(6) **Height of the wall:** The difference between the heights of the walls constructed in a day should not be more than 1 m. This limitation is suggested to prevent the differential settlement of the foundation.

(7) **Watering:** After the construction, the wall should be watered for time until the mortar sets properly. For the cement mortar, the watering is done for 7 days to 14 days while for lime mortar the watering is done for 14 days to 21 days.

3.14 COMPARISON BETWEEN STONE AND BRICK MASONRY

Stone masonry and brick masonry both are used commonly throughout the world. The different types of masonry have different types of advantages and limitations. The stone and brick masonry can be compared as follows:

(1) The stones are available naturally while the bricks are manufactured in the plants.

(2) The stones require dressing and modification in their raw form. The edges and sides are to be dressed properly in stone masonry while the brocks can be used as it is. There is no need of dressing in the case of bricks.

(3) The size of the bricks remains uniform throughout the construction while the size of stone masonry is non-uniform.

(4) The weight of the stones is more as compared to the bricks. So, the handling and transportation of the stones are complex while the bricks can be transported easily.

(5) The minimum thickness of a wall for stone masonry is 35 cm while the minimum thickness of brick masonry is 10 cm.

(6) The amount of mortar required in the stone masonry is more than the brick masonry due to the irregularity of the shape. The lining of the mortar remains thicker in the case of stone masonry.

(7) Stone masonry is more durable and it can withstand the action of external agencies in a better way as compared to brick masonry.

(8) The brick masonry offers more resistance to fire.

(9) Stone masonry doesn't require plastering while brick masonry requires a surface finish and plastering.

(10) The sewage, urine, human excreta, etc., damage the brickwork. So, brick masonry can't be used where it can come in contact with such types of waste.

(11) The bricks emit more heat as compared to the stone masonry.

3.15 TOOLS AND PLANTS FOR STONE AND BRICK MASONRY CONSTRUCTION

There is a number of equipment used for different purposes in stone and brick masonry. Figure 3.31 shows all the tools used for the masonry works. The bar code can be scanned for further information about the tools used in masonry works. The various tools that are used in the construction of stone and brick masonry are explained as follows:

(1) **Trowel:** It is used for lifting and placing the mortar. The trowel is made of steel while the handle of the trowel may be made of wood or plastic. It can be pointed or nosed from the front.

(2) **Pick axe:** Pick axe is used for digging and dressing stones.

(3) **Thick hammer:** The thick hammer is used for breaking stones. It can also be used for the rough dressing of the stones.

(4) **Mason square:** It is used for setting out the right angles during construction.

(5) **Chisel:** The chisels are used for the dressing of stones. Chisels are available in many sizes depending on their functions.

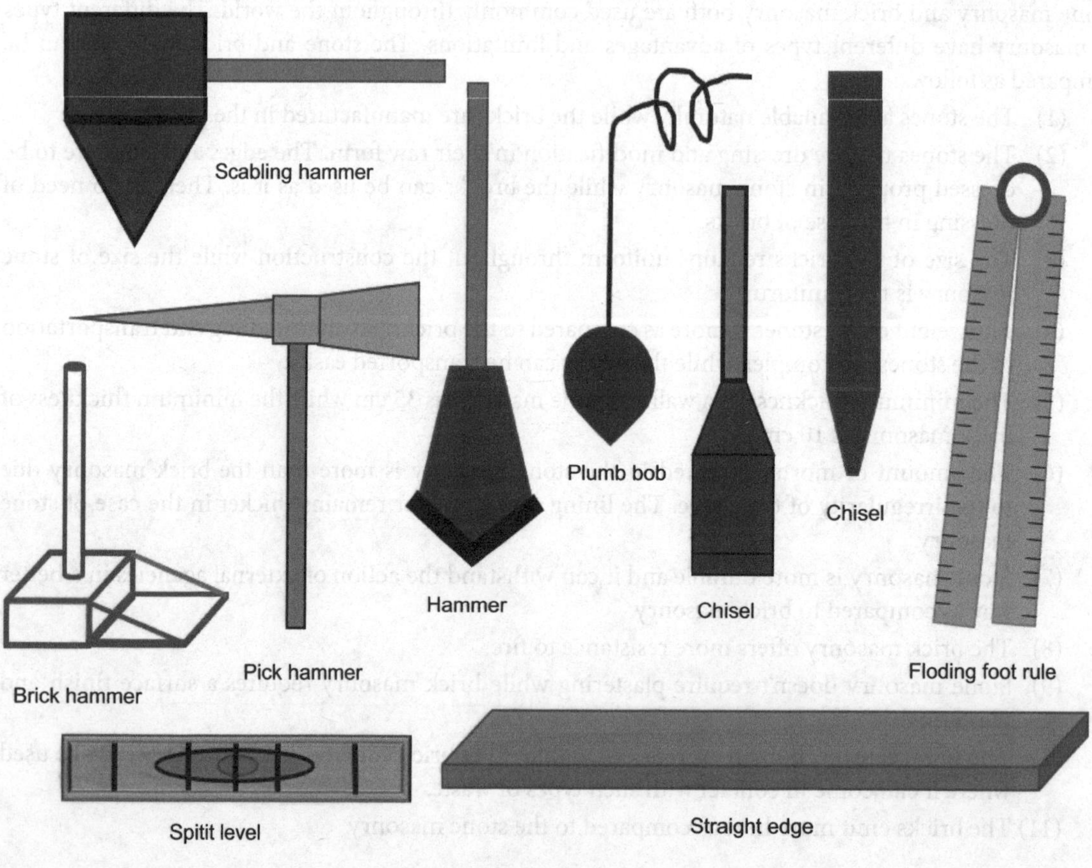

Fig. 3.31: Tools in masonry work

(6) **Plumb bob:** Plumb bob is used for checking the verticality of the edges.

(7) **Spirit level:** Spirit level is used for checking the horizontality of the surface.

(8) **Brick hammer:** Brick hammer is sharp at the face. It is used to cut the brick and masonry in the desired dimension.

(9) **Wire brush:** The wire brush is used for the removal of stone chips and excess mortar in brick masonry.

(10) **Folding foot rule:** The folding foot rule is used for the measurement of the wall. It can be folded.

(11) **Straight edge:** The straight edge is used for smoothing the plastering work.

3.16 HOLLOW CONCRETE BLOCK MASONRY

The concrete blocks are made of ordinary Portland cement and aggregate. High-density blocks are fabricated using cement, sand, and gravel. The concrete blocks can be categorized in solid concrete blocks and hollow concrete blocks. In the solid concrete blocks, the cavity is not provided or remains below 25% of the gross cross-sectional area. In the hollow concrete blocks, one or more cavities are provided and it can be up to 50% of the total gross cross-sectional area. The advantages of the hollow concrete blocks are as follows:

- The hollow concrete blocks can be manufactured in large sections as shown in figure 3.32.
- These are light in weight due to the cavity. So, the handling and placing of the hollow concrete blocks are easy.
- The construction of structures with hollow concrete blocks is quick and easy. It consumes less time in construction.
- The buildings can be made earthquake-resistant using hollow concrete blocks. For this purpose, these blocks can be reinforced with steel bars and concrete.

Fig. 3.32: Hollow concrete block

- These blocks provide good acoustic and thermal insulation due to the availability of the cavity.
- The cavities in the hollow concrete blocks are used for the installation of electrical setup and plumbing systems.
- The production cost of the hollow concrete blocks is slightly lesser than the bricks. These blocks consume less mortar as compared to bricks.
- The maintenance cost of the hollow concrete blocks is lesser.
- There is no need for formwork like in the case of construction of concrete members.
- The hollow concrete blocks are divided into A and B grades. If the minimum density for the A-grade hollow concrete blocks is 1500 kg/m^3. For B-grade blocks, the density varies from 1100 kg/m^3 to 1500 kg/m^3.

3.17 COMPOSITE MASONRY

Composite masonry is composed of more than one type of masonry shown in figure 3.33. The construction materials like stones, bricks, etc. are used for improving the performance and appearance of the structure. The composite masonry is used in the following ways:

(1) **Stone facing with brick backing:** The wall is constructed of bricks and the stones are provided at the face. The stone facing with brick backing is provided to improve the appearance of the structure. The appearance of the structure is improved with less cost.

(2) **Stone slab facing with brick backing:** Stone slabs of 5 cm to 10 cm thickness are provided on the face of the wall while the wall is constructed of bricks. This technique also improves appearance with less cost.

(3) **Brick facing with concrete backing:** In this type of composite masonry, concrete is provided in the back for providing strength while the bricks are used on the surface.

(4) **Ashlar facing with brick backing:** Ashlar masonry is used on the face while the wall is constructed of bricks. The stones used can be roughly dressed.

(5) **Ashlar facing with rubble backing:** The roughly tooled ashlar masonry is provided on the face while rubble masonry is used as backing in this type of composite masonry. The rubbles are less costly and the construction of this composite masonry proves economical.

Fig. 3.33: Representation of composite masonry

Composite masonry imparts higher strength and durability. The appearance of the structures can be improved using composite masonry. The optimum use of the locally available material helps in achieving economy in construction.

However, composite masonry is high in cost and requires skilled workers due to the complexity of construction. Glass composite masonry is also used nowadays. The barcode provided at the top of this section can be scanned to know more about the composite masonry.

3.18 SCAFFOLDING AND ITS PURPOSE

Scaffolds are temporary structures constructed for supporting the workers and the materials. The scaffolds are platform-like structures that are used for structures with a height of more than 1.5 m. The scaffolds are removed once the construction of the building component is over.

The failure of scaffoldings is life-threatening for the works. So, the scaffolds should be stable in adverse environmental conditions like rain, storm, etc. The method of construction of the scaffolds is known as scaffolding. The scaffolding is used for the following two purposes:

(1) Scaffolding provides a place for the worker or labourer so that they can lift up the construction. The artesian can work up to some limited height without support. Scaffolding provides elevated space for extending the work to the required height.

(2) Scaffolding provides space for keeping the material near the artesian at elevation.

3.19 TYPES OF SCAFFOLD

The scaffolds are constructed using bamboo poles, ropes, boards, etc. The members used in the construction of the scaffolds are as follows:

(1) Standards are the vertical members. These are supported on the ground or penetrated into the ground.

(2) Ledgers are the horizontally running members and remain parallel to the wall. These are tied to the standards.

(3) Braces are the diagonal members and these are fixed on the standards.

(4) Putlogs are the transverse members that remain perpendicular to the surface of the wall. These are supported by the wall at one end and the other end is supported by ledgers.

(5) Scaffold boards are placed horizontally to place the material and support the workers.

The different types of scaffolds have their own suitability and limitations. The selection of a scaffold is based on the loading condition, field condition, type of work, and budget. The following types of scaffolds are used based on the different types of masonry work and the shape of the wall:

(1) Single scaffold
(2) Double scaffold
(3) Cantilever scaffold
(4) Suspended scaffold
(5) Trestle scaffold
(6) Steel or tubular scaffolds

3.19.1 Single Scaffold

The single scaffold is also known as a brick masonry scaffold. One end of the scaffold rests in the holes of the walls and the other end is supported by the standards. The standards are penetrated into the ground. As the scaffold is supported by the standards on only one side, it is known as a single scaffold. Figure 3.34 shows the single scaffoldings. The following points are considered during the construction of a single scaffold:

- Single scaffolds are used for ordinary buildings.
- The components of the scaffolding are arranged parallel to the wall at a distance of around 1.2 m. The standards are placed at a distance of 2 m to 2.5 m.

- The ledgers are connected to the standards at a spacing of 1.2 m to 1.5 m.
- The putlogs are placed at a spacing of 1.2 m to 1.5 m.

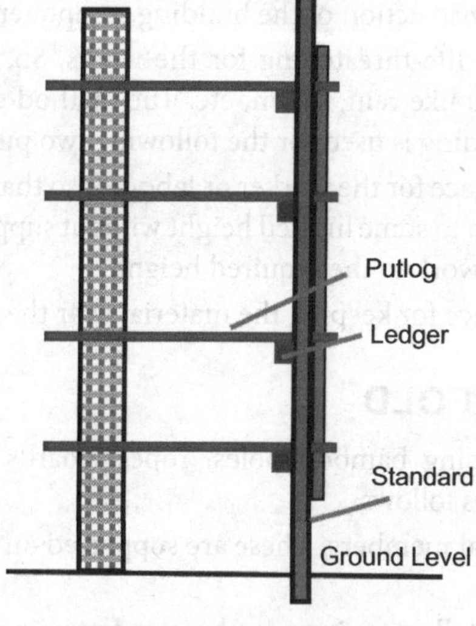

Fig. 3.34: Single scaffolding

3.19.2 Double Scaffold

It is very difficult to provide holes in the wall for inserting the putlog. So, a double scaffold is provided which is stronger in nature. It is also known as mason's scaffold. In double scaffolding, two rows of scaffolds are used. The first row of the scaffolds is placed at 20 cm to 30 cm from the wall while second row is placed at 1 m away from the wall. The scaffolding is given more support by rakers. Figure 3.35 shows the double scaffolding.

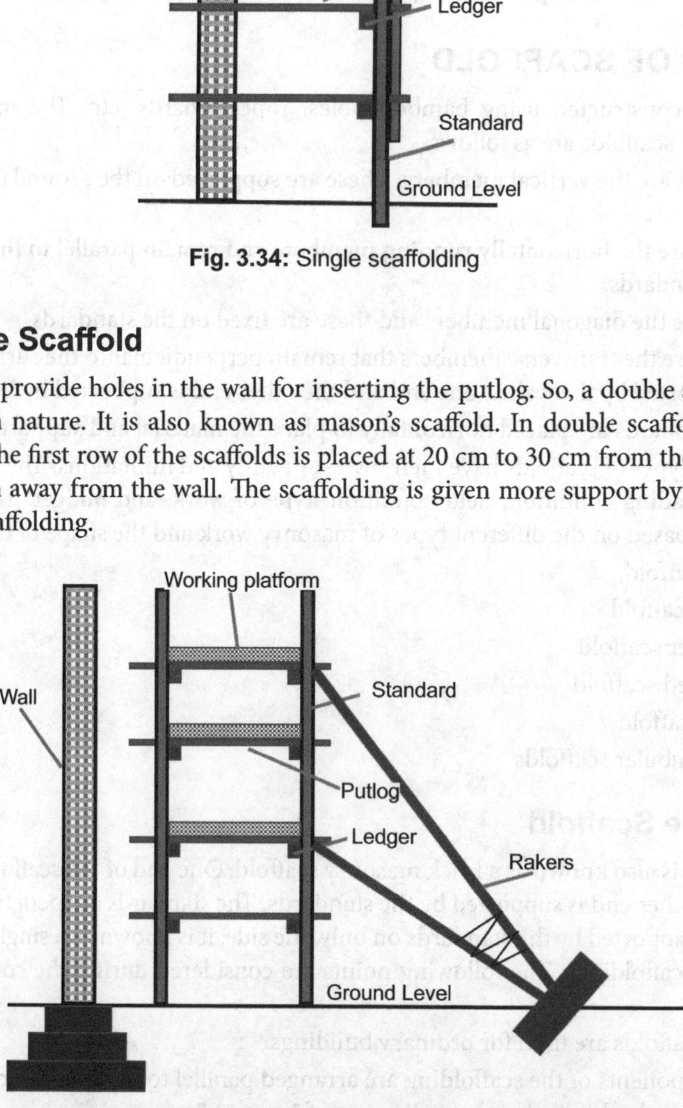

Fig. 3.35: Double scaffolding

3.19.3 Cantilever Scaffolding

A cantilever scaffolding is also known as needle scaffolding. The cantilever scaffolding is used when the ground is weak and can't support the standards. The cantilever scaffolding is used when the standards can't be supported on the ground due to some obstruction or traffic. The needle supports the scaffold structure and it is braced. The cantilever scaffolding can be of single frame or two frame type as shown in figures 3.36 (a) and 3.36 (b) respectively.

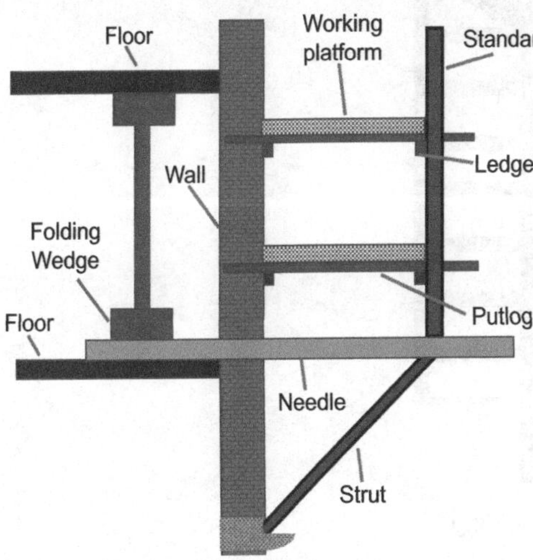

Fig. 3.36 (a): Single frame cantilever

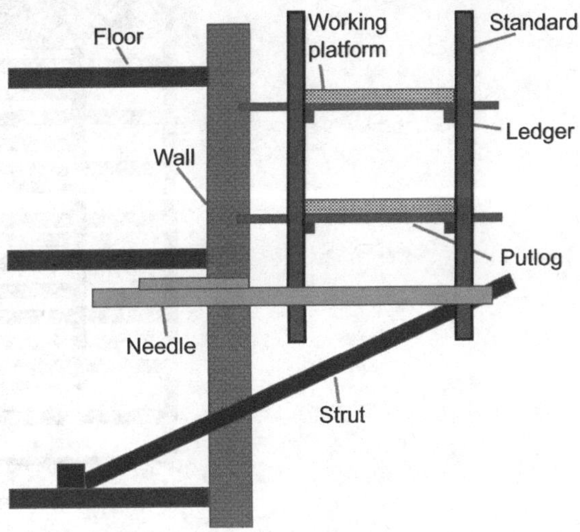

Fig. 3.36 (b): Double frame cantilever

A single frame cantilever scaffold is supported by a wall at one end and standards by another end. The needle passes through the wall and the strut is supported at the base of the wall. In a double frame cantilever scaffold, the strut is supported on the floor due to the weakness of the ground.

3.19.4 Suspended Scaffolding

The suspended scaffolding is used for small works like painting and finishing the walls. The working platform is suspended in air. The platform is supported by chains and ropes at the roof. The platform provides flexibility as it can be adjusted at any level with the help of ropes or chains. This type of platform is light weighted. If the elevation is high, the suspended scaffolds are the most economical solution. The suspended scaffolds can be fixed to the trusses on the roof or these can be suspended by pulleys. The pulley or crane-based suspended scaffold can be more flexible which can vary the level of platform based on their requirement. Figure 3.37 shows the suspended scaffolding.

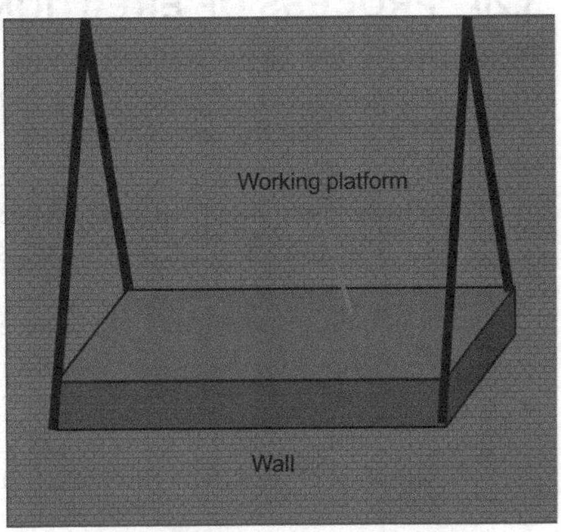

Fig. 3.37: Suspended scaffolding

3.19.5 Trestle Scaffolding

Trestle scaffolding is also known as ladder scaffolding. The working platform is mounted on the tripods or ladders. These tripods or ladders may move with the help of wheels. Such types of the scaffold are suitable for only up to 5 m in height and can be used for light works. These are used for painting and small repair purposes. Figure 3.38 shows the trestle scaffolding.

Fig. 3.38: Trestle scaffolding

3.19.6 Steel Scaffolding

The steel scaffolding is the same as timber scaffolding and may have similar arrangements as discussed in previous sub-sections but the members of the scaffolding are made of steel. Steel gives more tensile strength and can take higher loads. Steel tubes are used instead of timber in this type of scaffolding.

3.20 PROCESS OF ERECTION OF SCAFFOLDING

The correct erection of the scaffolds is very important for safe construction. The scaffolds should be erected carefully and the work should be conducted under the supervision of the experts. The supervisor is responsible for checking and maintenance of the scaffolds. The following procedure is followed for the erection of scaffolds:

(1) The first step is to prepare the foundation at which the scaffolds are to be supported. Weak soil or mud is replaced by gravel. The ground is excavated instead of backfilling. If the soil is weak, cantilever scaffolding is selected. The cantilever scaffolds are supported on the floor through a needle and strut arrangement. The soil should not be disturbed excessively.

(2) The equipment for the erection of the scaffolds must be placed near the place of work. The equipment should be unloaded in the pattern of use.

SCAN ME

(3) The frames are erected safely. The elements of the scaffolds like ledgers and braces are properly connected. The joints are properly locked and inspected. The elements of the scaffolds should not have deteriorated.
(4) The erection of the frames must be under the strict supervision of the experts. The supervisor must check the scaffolds for proper jointing.
(5) The alignment of the frames should be checked. The alignment must be proper for each level of the scaffold.
(6) The putlogs are placed on the ledgers and joined properly. The joints are inspected by the supervisor.
(7) In the case of suspended scaffolds, the chains, pulleys, etc. are planned and inspected by the supervisor.
(8) After the erection of the scaffolds, a final inspection is conducted. The scaffolds require daily inspection due to safety reasons.
(9) The supports under each leg should be inspected. If the level of support is not proper, action should be taken.
(10) Frames should be checked for alignment using the plumb bob.
(11) All the ties and braces must be inspected for locking.

3.21 DISMANTLING OF SCAFFOLDING

As it is mentioned that the scaffolds are temporary structures, so these are to be removed when the work is completed. The dismantling of scaffolds involves equal risk as the erection of the scaffolds. The dismantling of scaffolding is done as follows:

(1) The scaffolding is checked for structural stability. If it seems unsafe at any part, that part is reconstructed before beginning the dismantling procedure.
(2) The dismantling is done from top to down.
(3) The workers should work at least on the two planks placed in tier of frames.
(4) The ties should not be removed until the level of scaffolding to which they are attached is reached. The planking should be moved downwards as the dismantling progresses.
(5) The workers should not climb the standards or braces. They should remained inside the scaffolding.
(6) The fastening items should be removed from the bottom of the frames.
(7) The scaffolding components are lowered carefully without dropping them roughly.

3.22 SHORING AND ITS PURPOSE

Shoring may be defined as the temporary supports provided to the structures or excavation. The lateral supports are provided to support the walls or excavation. In the case of differential settlement or bulging out of walls, the shoring can be used during repair works. The shoring is provided for the following purposes:

(1) If the wall swells or bulge outside, the shoring can be provided. The bulging out of the wall can occur due to the poor workmanship.
(2) If the foundation is settled unequally, the cracks are induced in the walls. Shoring can be provided during the repair work of the cracked wall at the time of differential settlement of foundation.

(3) If demolition works in nearby structures are going on, shoring can be provided.
(4) When underlying supports are to be removed from the walls, these can be supported for prevention of the failure.
(5) If any wall is prone to failure or becomes unstable due to geological conditions, the shoring can be provided.
(6) The shoring increases the safety during work. So, the workers can efficiently work in safe environment.

3.23 TYPES OF SHORING

The different types of shoring have different functions. There are the following three types of shoring in construction:

- Dead shoring
- Raking shoring
- Flying shoring

3.23.1 Dead Shoring

In dead shoring, the support to the building components is provided by the horizontal members like needle. The dead shoring supports the dead load that acts vertically downwards. It is normally used for transfer of the load from the wall through horizontal beam resting on the two vertical members. If the wall is weak or defective at the bottom part, the dead shoring is used to transfer the load of the wall to the ground. The dead shoring arrangement is shown in the figure 3.39.

The wall under the needle can be demolished or removed as the whole load above the needle's level is taken by the needle. The needle transfers the load of the components to the dead shores. These dead shores transfer the load to the floor and ground. The needles are arranged at 1 m to 2 m spacing. The needles can be braced to strengthen the support. The floors of the building are supported properly. As the load on the shores can't be predicted accurately, so the factor of safety is kept high.

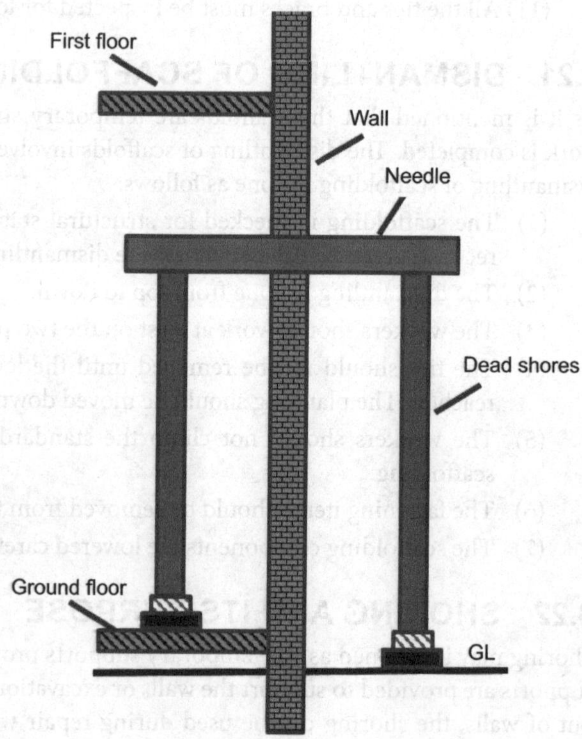

Fig. 3.39: Dead shoring

3.23.2 Raking Shoring

In the raking shoring, the support to the wall is provided by inclined members known as rakers. The wall plate is attached to the wall with the help of needle. The needle is inserted 15 cm deep in the wall. Cleat is provided to increase the support to the needle. This whole arrangement is supported by the rakers which are braced. The bottom of the rakers is connected to the sole plate. The rakers are inclined at the angle of 45° ideally but the inclination angle can be extended up to 70 ° but in no case it should be more than 75°. The soleplate should be sufficiently large to accommodate the cleat and raker. The factor of safety is kept high as the load on the rakers can't be calculated accurately. The rakers are properly braced for increasing their strength. Figure 3.40 shows the raking shoring.

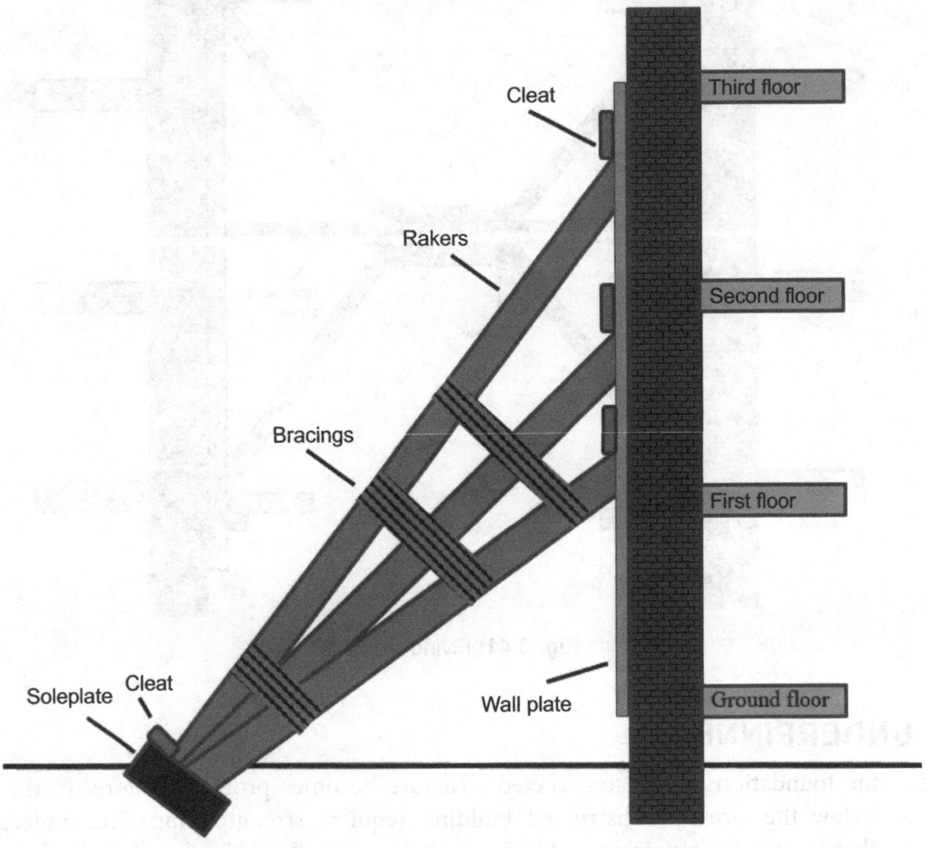

Fig. 3.40: Raking shoring

3.23.3 Flying Shoring

The flying shoring is provided to the wall of two buildings when the intermediate building is to be demolished. It prevents damage to the adjacent structures during the process of collapse of the intermediate building. The shoring system doesn't reach the ground, it is known as flying shoring. The walls are secured by the wall plates and horizontal and inclined struts are used to support the wall. When there is a significant distance between the two walls, a horizontal beam remains unsafe to use. In this case, a horizontal truss system is used for better stability. The horizontal shores are placed at a spacing of 3 m to 4.5 m along the walls. The bracings can be provided. Single shores are only suitable for a maximum distance of 9 m between the two walls. If the distance between the walls is more than 9 m, truss system is preferred. The struts are inclined at 45°. Figure 3.41 shows the flying shoring.

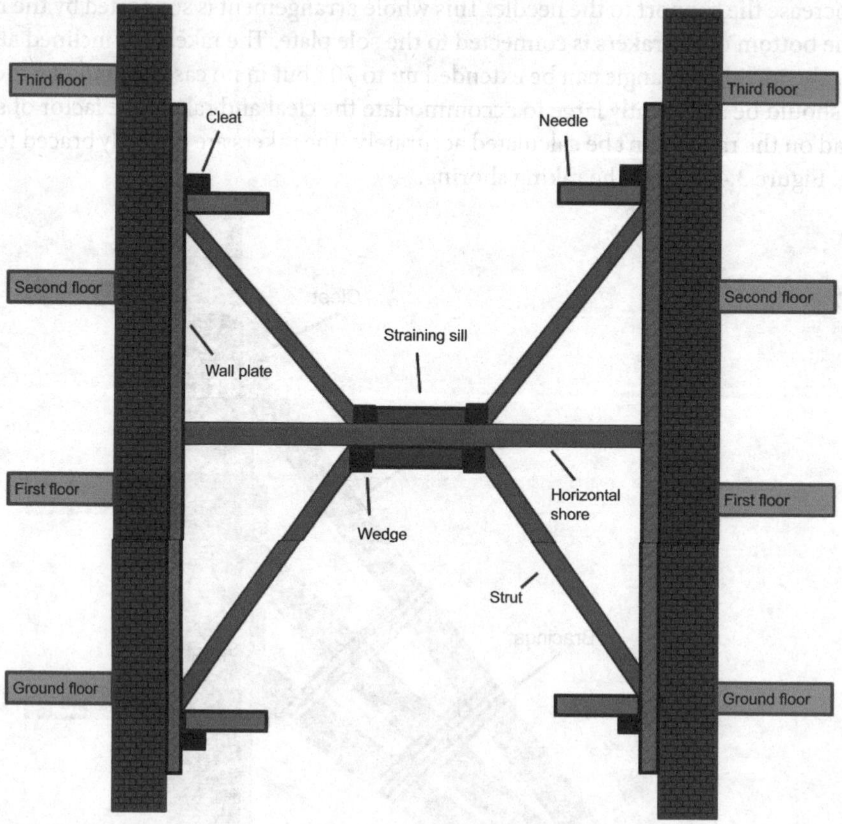

Fig. 3.41: Flying shoring

3.24 UNDERPINNING

Sometimes the foundation of the constructed structure becomes prone to failure. In this case, the foundation below the already constructed building requires strengthening. The underpinning is the process that is used for repairing and enhancing the strength of the foundation of structures. A temporary structure is installed at the additional depth below the foundation to increase the strength of the foundation and sufficient bearing capacity is achieved.

The underpinning work for the building is decided by carefully examining the behaviour of the existing structure. If the structure shows excess settlement and chances of failure, underpinning can be suggested. The underpinning may be suggested in the following conditions:

(1) If the functions of a building are changed, the load can be increased on the foundation. For example, if a normal building is converted for the operation of heavy machinery, the vibration loads can induce. In this case, the foundation of the existing structure may not have sufficient strength. Underpinning may be suggested for the strengthening of the existing foundation in this case.

(2) The underpinning can be suggested for the protection work. If the existing foundation is not sufficiently strong, it may require strengthening. If there is excess excavation of the soil in the vicinity of the foundation, the bearing capacity may decrease.

(3) If the foundation is to be protected from external agencies like wind and piping action, the underpinning can be proposed.

(4) If the construction of a basement is proposed for the existing structure, the underpinning is required.

(5) If there are errors in the design of the foundation of the structure, the existing building can become prone to failure. In this case, the design can be corrected by enhancing the area of the foundation.

(6) The fluctuations of the water table may decrease the bearing capacity of the soil. Sometimes, the timber piles can be deteriorated by the action of variation of water table or insects. In this case, the underpinning is required for strengthening the foundation. The barcode can be scanned to know more about the underpinning.

SCAN ME

3.25 DEFINITION OF FORMWORK

The concrete can be cast in any desired shape but at the time of pouring concrete has very less shear strength. So, during the time poring it requires support to maintain the desired shape. Formwork may be defined as the temporary support to keep the concrete in its place. This temporary support is provided until the concrete achieves sufficient strength to bear its own weight. The formwork is kept strong enough to sustain the load of the wet concrete.

At the time of removal of formwork, it should be able to withstand the load of labourer. If the formwork is deformed during the setting of the concrete, the shape of the structural member is also deformed which is very difficult to improve later.

3.26 REQUIREMENTS OF FORMWORK

The formwork plays an important role in the construction of the concrete members in the desired shape. The formwork should have the following characteristics:

(1) **Ease of removal:** A good formwork is always easy to remove. If the formwork requires excess hammering for removal, it may damage the corners and edges of the structural member. So, the formwork should be designed in a way that it can be removed easily with minimum hammering.

(2) **Dimensions:** The dimensions of the formwork should be according to the structural components of the building. The dimensions are carefully considered and the internal dimension should be kept in mind while designing the formworks. The shape and dimension of the formwork can be planned from the drawings and plans of the building.

(3) **Economy:** The formwork should be cost-effective. The purpose of the formwork is just to keep the concrete in shape and it doesn't contribute to the stability and strength of the structure. So, the design of the formwork should be economical. The cost of the formworks can be 20% to 30% of the total cost of the concrete structure. The cost may vary based on the complexity of the structure and can reach even up to 60% of the total cost of the concrete structure.

(4) **Rigidity:** The formworks should be able to withstand the deflections. These should be sufficiently rigid to face the deflections induces. The maximum deflection in formworks for the surface work is 1/300 of the span while for the hidden surfaces it is 1/150 of the deflection.

(5) **Strength and durability:** The formworks should be sufficiently strong to withstand the loads. They should be able to bear the load of the poured and wet concrete. These should be strong enough to take the impact of the hammers and load of a person while removing them. The formworks are reused for other works. So, these should be able to withstand the impact of weather and temperature.

(6) **Reusability:** As discussed before, the formworks are reused many times. So, the formworks should be reusable in nature. They should not show variation in their characteristics after using those one or two times. They can be used multiple times.

(7) **Leakage:** The joints of the formwork should be tight to prevent the leakage of the water. It is evident that the concrete mix is designed for particular water content. If the water from the formwork leaks in excess, it can vary the properties of the concrete itself.

(8) **Finish of surface:** The inner surface of the formwork should be smooth. The oiling of the inner surface is done for achieving a smooth surface.

3.27 MATERIALS USED IN FORMWORK

A number of materials are used for making the formworks. The following material is used for making the formworks:

- Timber
- Plywood
- Steel
- Plastic
- Aluminium

Timber is the most commonly used material for formworks. It is easily available and is used for a long time. The prefabricated formwork of aluminium and steel is also used. Steel imparts high strength while aluminium is light in weight with good strength.

3.28 TYPES OF FORMWORK

The formworks are classified based on the material. As discussed in the previous section, different types of material can be used for formworks. The materials for the formwork have different characteristics and their own advantages and disadvantages.

3.28.1 Timber Formwork

Timber is used most of the time for formworks. It is easily available and can be cut into any desired shape. The timber used for the formwork should be free from any defects. There should not be any knot in the timber. Softwood is generally used for making formworks. The hardwood is very difficult to cut and also it resists the nailing process. If the wood is excessively dry, it creates an uneven surface. So, a partially treated softwood is preferred for formworks. Pine, kail, cedar, etc. can be used for making formworks. Timber as a material for the formwork offers the following advantages:

(1) The timber shuttering is light in weight. It can be transported and placed easily.

(2) Timber formworks are easy to install and remove.

(3) It is easy to cut timber in any desired shape.

(4) The timber for formwork is easily available as timber is a universal material.
(5) The timber formwork is economical for small projects.

The timber formwork has less durability as compared to other types of formwork. The termite attack is common on the timber formwork. So, the coatings for termite attack prevention are provided. The surface of the timber is not as smooth as plastic and steel.

3.28.2 Plywood Formwork

The plywood is made by gluing the sheets of plies in multiple numbers. Plywood is the advanced version of timber. The durability of plywood is more than timber and the plywood formwork can be used more times as compared to timber formwork. The plywood formwork offers the following advantages:

(1) The surface of the plywood is very smooth as compared to the timber formwork. So, the members constructed using plywood formwork show a very smooth surface finish.
(2) The durability of the plywood formwork is higher than timber formwork.
(3) The plywood formworks can be installed easily as compared to the steel formworks

The weight of the plywood formwork is more than timber formwork while these are lighter than steel formwork. These require care at the time of installation and removal as these can get damaged by hammering.

3.28.3 Steel Formwork

The fabricated steel sections are used for making the steel formwork. Steel formwork is installed using bolts and clamps. The initial cost of the steel formwork is high but in long run these are economical. The major advantages of the steel formwork are as follows:

(1) The strength of steel formwork is very high. It can take heavy loads.
(2) The dimensions of the steel formwork are not affected by moisture or water. These don't shrink or expand due to the action of water.
(3) The surface of the steel formwork remains smooth.
(4) Steel formwork is highly durable. They can be used in more repetitions as compared to other types of formwork. They can be used more than 100 times.
(5) The steel formwork can be made watertight and the loss of water during formwork can be avoided using steel formwork.

The steel formworks are expensive comparatively. If the small nuts or bolts are missing, the formwork gets delayed. The major problem with the steel formwork is corrosion. So, treatment for the prevention of corrosion is required.

3.28.4 Plastic Formwork

Plastic formwork are made from good quality plastic with an interlocking facility. These type of formwork are light in weight and can be used multiple times. These can be fabricated in any desired shape. The plastic formwork offers the following advantages:

(1) The handling cost of the plastic formwork is less due to its lightweight. The transportation and placing of the plastic formwork are easy.
(2) Plastic formwork is easy to install but should be installed carefully.

(3) The durability of plastic formwork is good. It is not affected by water or air.

(4) These are economical due to their repetitive usage.

3.28.5 Aluminium Formwork

In aluminium formwork, fabricated sections of aluminium are used. The advantages of aluminium formwork are the same as steel formwork. It is lighter in weight as compared to steel formwork. The strength of the aluminium is sufficient to take the heavy loads and these can be fabricated in any desired shape.

3.29 REMOVAL OF FORMWORK

The removal of formwork is also known as the stripping of formwork. The formwork should be removed when the concrete gains enough strength to withstand its self-weight and other loads coming to it. The removal time of the formwork depends upon the time taken by the concrete for hardening. The hardening time depends on the type of cement used and the temperature. The structural members like the beam take live load during the construction too. So, care should be taken for such members before the removal of the formwork.

The strength of the members should be tested using concrete cubes before the removal of the formwork. The formwork should be removed when the concrete gains sufficient strength. The concrete members must not collapse or deflect after the removal of formwork. If there are problems of freezing and thawing, the removal of formwork is delayed. The following points should be considered before removing the formwork:

- The formwork should be removed only if the concrete has gained sufficient strength to take the loads coming to it. The hardening of the concrete depends upon the type of cement, temperature, moisture, grade of concrete, etc. So, these factors should be taken care of before the removal of the formwork.
- The structural members must not deflect or collapse after the removal of the formwork. If the members are deformed, these can be repaired instantly.
- The cow bars should not be forced during the removal of the formwork because it can damage the member.
- The removal of formwork must be done under the supervision of a qualified expert.

If ordinary Portland cement (OPC) is used for the construction, the formwork of the walls, columns, and vertical faces of beams can be removed between 24 hours to 48 hours. The formwork of slabs for which props are left can be removed in 3 days while for the beams with props left, it can be 7 days.

UNIT SUMMARY

- Stone masonry consists of stones and mortar while brick masonry consists of bricks and mortar.
- Stone masonry is classified into rubble masonry and ashlar masonry.
- Roughly dressed and irregular-shaped stones are used in rubble masonry while the stones used in ashlar masonry are relatively well dressed.
- Stone masonry can be further subdivided depending on its arrangement.

- The stones are connected in different ways. The connection between the stones is known as a joint.
- The absorption of water during 24 hours of immersion of stone should not be more than 5 %.
- Brick masonry can be arranged in many ways. The arrangement of stone masonry is known as a bond.
- The stretchers are placed back to back in stretcher bond while if the headers are placed at the facing, it is known as header bond.
- English bond is stronger than the Flemish bond while the Flemish bond is aesthetically pleasing.
- Bricks should be uniform in size and should be durable.
- If the walls meet at a point at 90°, it is called a Tee junction.
- If the walls cross each other at 90°, it is called a cross junction.
- If the walls meet at any angle, it is known as a squint junction.
- The bricks should be soaked for at least 12 hours before using them in construction.
- The hollow concrete blocks are made of OPC and aggregates.
- The hollow concrete blocks are economical in nature and can be manufactured in large sections.
- The maintenance cost of the hollow concrete boxes is less and there is no requirement for finishing and painting.
- When two or more types of masonry are used together for achieving the economy is known as composite masonry.
- Scaffolds are the temporary structures that are used to support the material and artesian. If the height of the wall is more than 1.5 m, scaffolds are required.
- Single scaffolds are supported by a wall on one side and another side is supported by standards. Single scaffolds are used for ordinary construction.
- The double scaffold is supported on both sides by standards.
- Scaffolds are erected and dismantled in the presence of expert.
- The scaffolds should be erected and dismantled very carefully as there is a huge risk involved for the workers.
- Shoring is the temporary structure to support the excavations and structures.
- Dead shoring supports the weight of the structure on it and the wall below the dead shoring can be dismantled and constructed again.
- The underpinning is the process that is used for repairing and enhancing the strength of the foundation of structures.
- Formwork may be defined as the temporary support to keep the concrete in its place. This temporary support is provided until the concrete achieves sufficient strength to bear its own weight and other loads coming to it.
- The formwork should be removed only if the concrete has gained sufficient strength to take the loads coming to it.
- The hardening of the concrete depends upon the type of cement, temperature, moisture, grade of concrete, etc.
- If ordinary Portland cement (OPC) is used for the construction, the formwork of the walls, columns, and vertical faces of beams can be removed between 24 hours to 48 hours.

EXERCISES

Multiple Choice Questions

1. Granite belongs to the rock group:
   (c) Igneous
   (b) Metamorphic
   (c) Sedimentary
   (d) None of the above
2. Shale belongs to the rock group:
   (d) Igneous
   (b) Metamorphic
   (c) Sedimentary
   (d) None of the above
3. The heavy stones are arranged one over another and a retaining wall is constructed without using mortar. Identify the type of stone masonry.
   (c) Random rubble
   (b) Dry rubble
   (c) Polygon walling
   (d) Flint walling
4. In flint walling, the lacing is provided at an interval of:
   (a) 1 m to 2m
   (b) 2m to 4m
   (c) 4m to 5m
   (d) None
5. Being a civil engineer, if you have to choose a masonry for fire resistance, which one will you choose?
   (c) Stone masonry
   (b) Brick masonry
   (c) Both can be used
   (d) None
6. Rubble masonry can be classified into:
   (c) 4 types
   (b) 6 types
   (c) 8 types
   (d) 5 types
7. What is the modular size of the brick (in cm)?
   (c) 19*9*9
   (b) 20*10*5
   (c) 19*9*4.5
   (d) 20*10*10
8. Which bond is the strongest among the following?
   (c) Header bond
   (b) Stretcher bond
   (c) English bond
   (d) Flemish bond
9. Which bond is the most pleasant bond aesthetically among the following?
   (d) Header bond
   (b) Stretcher bond
   (c) English bond
   (d) Flemish bond
10. Which of the following is the advantage of hollow concrete block?
    (c) Sound insulation
    (b) Heat insulation
    (c) Light weight
    (d) All of the above
11. In a scaffold, the working platform is supported by wheels and can be moved. Identify the type of scaffolding:
    (a) Single scaffold
    (b) Trestle scaffold
    (c) Double scaffold
    (d) Cantilever scaffold

12. Name the arrangement that is provided to support the concrete after pouring.
    (a) Scaffolding (b) Formwork
    (c) Grouting (d) Underpinning

Answers of Multiple Choice Questions

1. (a), 2. (c), 3. (b), 4. (a), 5. (b), 6. (b), 7. (d), 8. (c), 9. (d), 10. (d), 11. (b), 12 (b)

Short and Long Answer Type Questions

1. Explain the different types of rocks. How are these formed?
2. What are the different types of stone masonry?
3. Write a brief note on rubble masonry.
4. Write a short note on ashlar chamfered masonry.
5. Explain the different types of joints in stone masonry.
6. What are the different types of closers? Draw neat sketches.
7. Draw a neat sketch of elevation and plan of 1.5 brick English bond.
8. Explain the Tee junction and its types in detail.
9. Explain the different types of squint junctions.
10. Compare the stone and brick masonry.
11. Being a civil engineer, what kind of precautions will you recommend during the construction of brick masonry?
12. Enlist the different tools used in stone masonry work. Also, write down their function.
13. Write a short note on cantilever scaffolding.
14. Explain the cantilever scaffolding with a neat diagram. What are the conditions in which you will recommend the cantilever scaffold?
15. Explain the different types of shoring.
16. Enlist the precautions to be taken during the erection and dismantling of the scaffolding.

KNOW MORE

The first glass block was patented by Gustave Falconnier in 1886. The glass bricks became a popular building material from 1930. These are aesthetically pleasing and attracted the architects. The glass bricks are used in the windows, building facades, etc. Maison Hermès in the Ginza District Tokyo is a fabulous example of using the glass as building masonry. La Maison de Verre in Paris was constructed in 1932 and is a house of glass. To know the history of glass block in buildings, scan the bar code.

SCAN ME

REFERENCES AND SUGGESTED READINGS

1. Chudley, R., Greeno, R. *Building Construction Handbook.* Elsevier, Butterworth-Heinemann, Amsterdam, 2007.
2. Punmia, B. C., et al. *Building Construction: An Elementary As Well As Advanced for Engineering Students.* Laxmi Publications, 2016.
3. Iano, J., Allen, E. *Fundamentals of Building Construction: Materials and Methods.* Wiley. 2019.
4. *National Building Code of India (Part IV), Fire and Life Safety, 2016*
5. Puybaret, E. *Sweet Home 3 D,* eTeks, 2016 (Open Access Software)

4 Building Communication and Ventilation

UNIT SPECIFICS

Through this unit we have discussed the following aspects:

- The different types of doors and their suitability.
- The different types of windows and their suitability.
- The different types of staircases and their suitability.
- The other means of vertical and horizontal communication in the building.

For helping the students in the visualization of the content and to enhance their practical knowledge, the figures are provided in 2D and 3D.

This unit contains short and long answer-type questions along with the multiple choice questions, a list of references, and suggested reading so that one can go through them for practice. Some QR codes are given which can be scanned for more information on various topics of interest. The QR code provides extra knowledge about important topics.

RATIONALE

This unit on building communication and ventilation gives the students an idea about the various facilities of communication provided in the building. It explains the different types of doors, ventilators, and windows using 2 D and 3 D diagrams. A brief description of elevators, lifts, and escalators is also provided. The different types of staircases and their significance is also discussed in the detail. The content provided in this unit is important for planners, architects, and engineers.

Horizontal and vertical communication is an important requirement of a building. The knowledge of horizontal and vertical communication is important for civil engineers and construction planners. The students can implement the knowledge gained from this unit in the planning, construction, and designing of the buildings.

PRE-REQUISITES

Basic knowledge of building components is required for studying this unit.

UNIT OUTCOMES

The list of outcomes of this unit is as follows:

U4-O1: Identify the suitable type of doors and windows for different types of buildings.

U4-O2: Differentiate between different types of the staircase and their suitability.

U4-O3: Explain the different types of fastenings and their suitability.

U4-O4: Understanding the standard sizes of windows and doors as per BIS standards.

Unit-4 Outcome	EXPECTED MAPPING WITH COURSE OUTCOMES (1 - Weak Correlation; 2 - Medium Correlation; 3 - Strong Correlation)				
	CO1	CO2	CO3	CO4	CO5
U4-O1	1	-	-	3	-
U4-O2	-	-	1	3	-
U4-O3	-	-	-	2	-
U4-O4	-	-	-	3	-

4.1 DOORS

A door may be defined as an open part in the wall, which is left for providing access to the persons in the building. Generally, doors are combination of different materials such as timber, steel, glass, etc. The internal parts of a building are connected using doors. A door basically comprises of two parts i.e. frame and a shutter. A frame is attached to the wall opening with the help of a horn and holdfast. The shutter or leaf is attached to the door frame.

4.2 COMPONENTS OF THE DOOR

A door is made of different components. The various components of the door offer different functions. Figure 4.1 shows the components of the door. The various components of the door are explained as follows:

- **Frame:** It consists of vertical members and horizontal members forming a structure, to which shutters are attached.
- **Shutters:** The shutter is the part of the door which is movable and can be opened. It consists of panels, rails, styles, etc. Shutters are fixed into the frame. It provides ventilation and light inside the room.
- **Head:** Horizontal member which is on the top or uppermost part of the frame is known as the head.
- **Horn:** Horizontal member which is projected outside the vertical member of the frame is known as the horn. It helps in fixing the frame into the openable part of the wall. The projected length is kept around10 to15 cm.
- **Style:** Vertical outside member of the shutter which helps in connecting the shutter to the frame with the help of the hinge.
- **Top rail:** Horizontal member present on the uppermost part of a shutter is known as a top rail.
- **Lock rail:** Horizontal member present in the middle of a shutter is known as a lock rail. This part of the rail consists of the locking arrangement of the door.
- **Bottom rail:** Horizontal member present on the lowermost part of a shutter.
- **Intermediate or cross-rail:** Additional horizontal members present in a shutter other than top and bottom rails. Horizontal members lie in between the top and the lock rail known as the frieze rail.

- **Panel:** An enclosed area between the adjacent rails and styles of a shutter.
- **Mullion:** Vertical member of a shutter, which helps in sub-divide the door shutter holdfasts is known as a mullion.
- **Holdfasts:** Holdfasts are made up of mild steel flat of section 30×6 mm. Z-shape is formed by bent the mild steel fast to fix or hold the frame into the openable part of the wall. Holdfasts are completely embedded into the wall opening. The horizontal length which is embedded inside the wall is kept around 20 cm.
- **Rebate:** Some depression or recess are made in the door frame to fix the door shutter. This depression or recess is known as a rebate.
- **Jamb:** Vertical face of the openable part of the wall which supports the door frame.

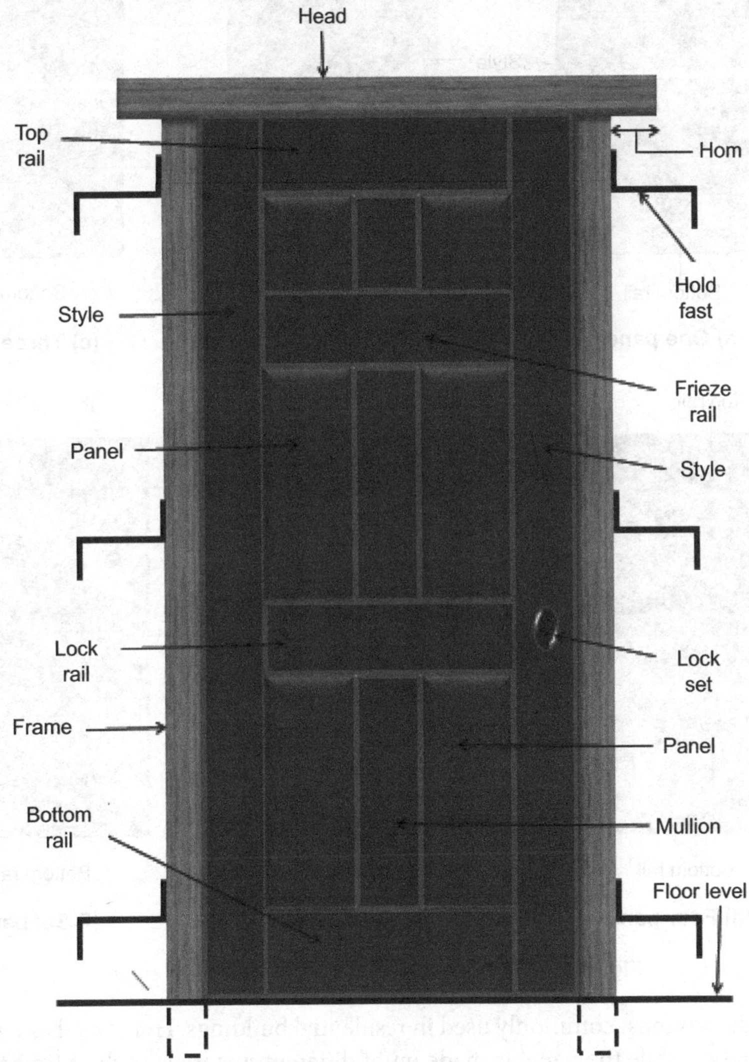

Fig. 4.1: Component of doors

4.3 PANELED DOORS

Full-paneled doors are made up of horizontal members known as rails and vertical members known as styles. The rails and styles are grooved from the inner side to receive the panel frame. Paneled doors have high strength and are aesthetically pleasing.

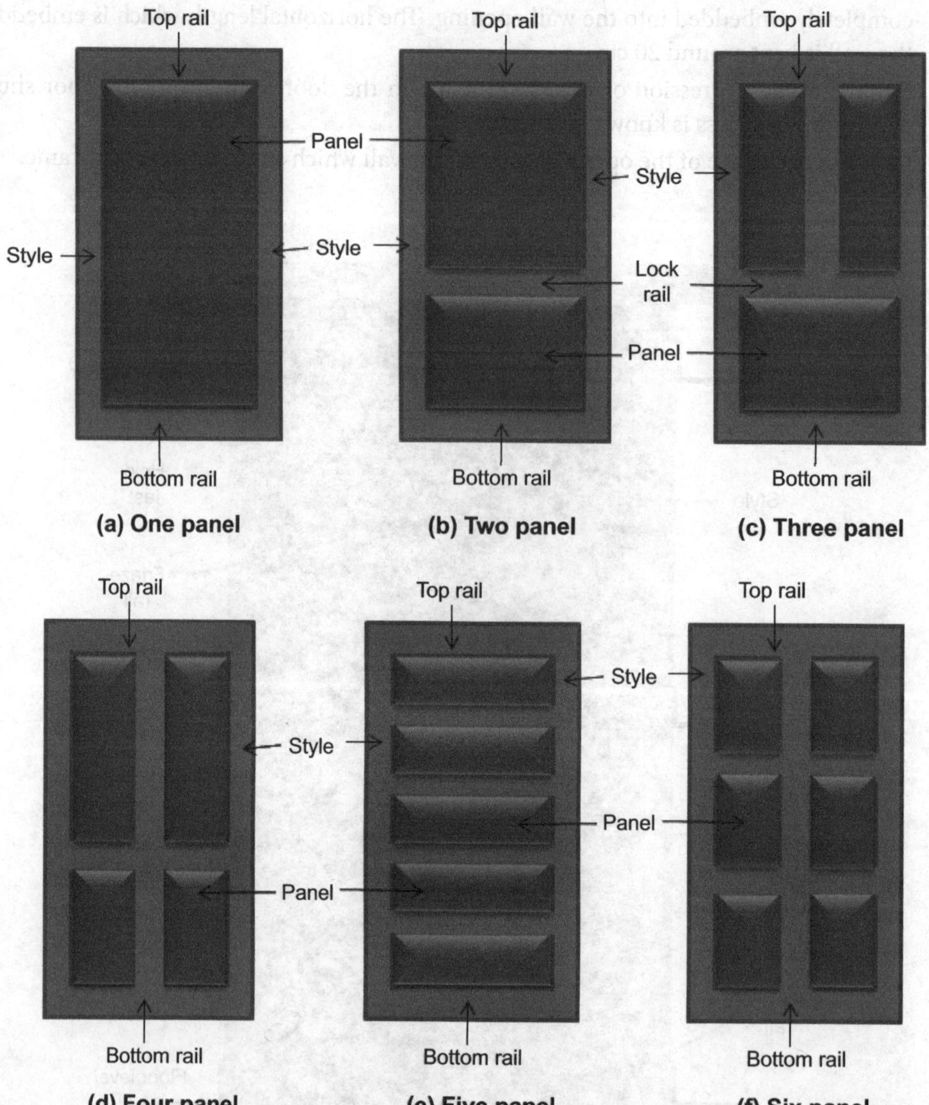

Fig. 4.2: Different types of single shutter panel door

This type of door is most commonly used in residential buildings. Horizontal and vertical members are made from timber while the panel is made up of different materials such as timber, plywood, glass sheets, etc. Different types of panel doors based on the number of panels (one, two, three, four, or multiple panels) are shown in figure 4.2.

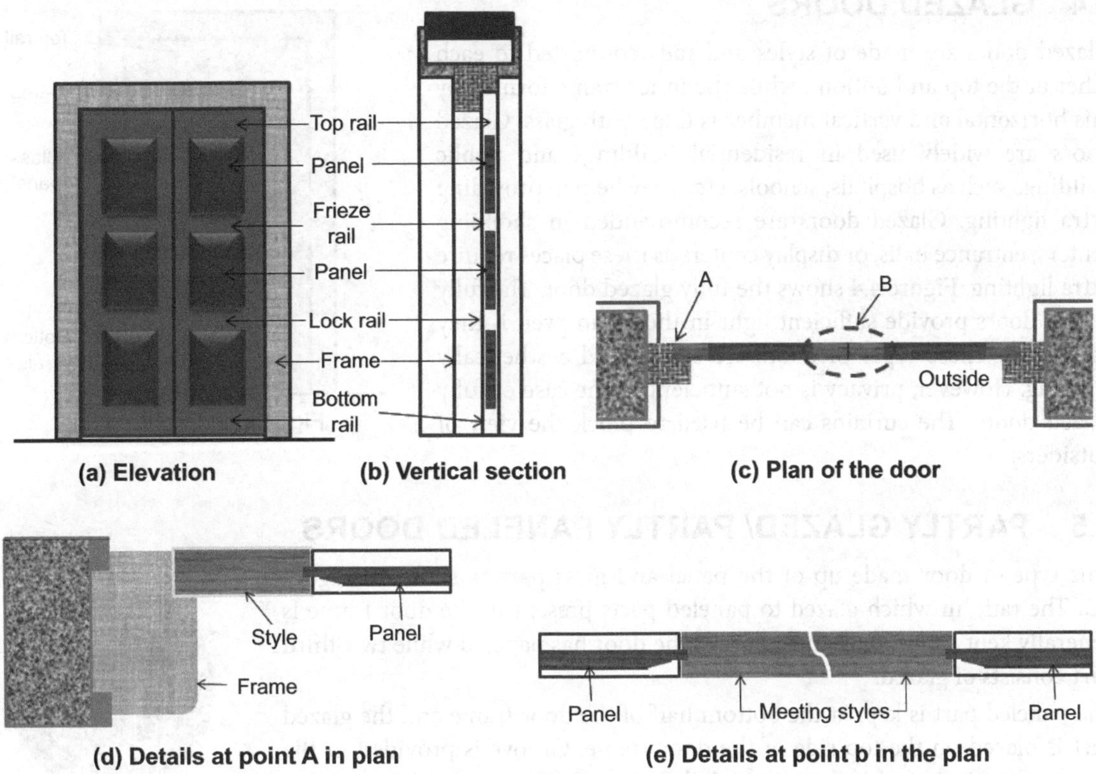

Fig. 4.3: Details of a double shutter panel door with six panel

When there is more than one panel in the door, intermediate or frieze rail and mullions are provided apart from the lock rail. Panel doors having a single shutter or leaf are generally used when the space of opening is small, while for wider openings, the door contains two shutters or leaves as shown in Figure 4.3. In a panel door having two leaves, each leaf is attached separately to the frame with the help of fixtures and fasteners. Some important points and details of panel doors are:

(1) Panel doors are made up of single pieces. The styles are used continuously from bottom to top without any joints in between them.

(2) The ends of the top rails, bottom rails, frieze rails, etc. are connected to styles.

(3) If any mullion is present, it is joined to adjacent rails.

(4) Intermediate rails and top rail are kept smaller than the bottom rail and lock rail.

(5) Where there is a lock rail provided in the panel, its center line kept 80 cm above the bottom rail.

(6) Style should have a minimum width of 10 cm.

(7) The lock rail and bottom rail have a minimum width of 15 cm.

(8) To improve the elevation of the door, extra wood bedding is provided on both sides of the panel.

(9) The wood panels should have a minimum width and thickness of 15 cm and 1.5 cm respectively.

(10) The wood panel as a single panel should not have an area greater than 5000 cm^2. This restriction only applies to the timber panel, and not to the plywood, glass sheets, etc.

4.4 GLAZED DOORS

Glazed doors are made of styles and rails connected to each other at the top and bottom, while the inner frame formed by this horizontal and vertical member is filled with glass. Glazed doors are widely used in residential buildings and public buildings such as hospitals, schools, etc. They help in providing extra lighting. Glazed doors are recommended in shopping centers, entrance halls, or display centers as these places require extra lighting. Figure 4.4 shows the fully glazed door. The fully glazed doors provide sufficient light in the room even if they are closed. These types of doors are stylish and aesthetically pleasing. However, privacy is not sufficient in the case of fully glazed doors. The curtains can be used to block the view of outsiders.

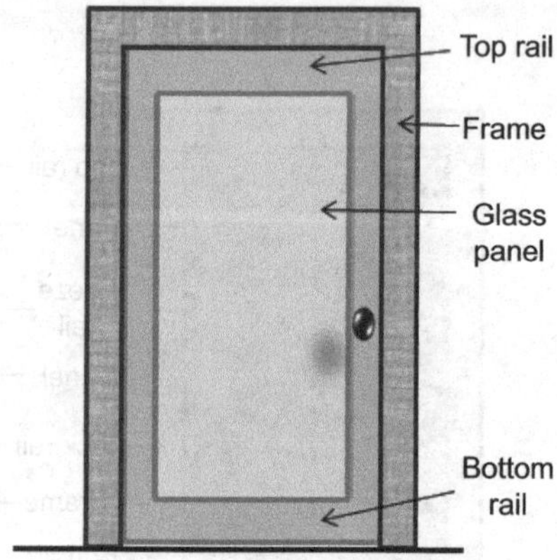

Fig. 4.4: Fully glazed door

4.5 PARTLY GLAZED/ PARTLY PANELED DOORS

This type of door made up of the panel and glass part as shown in figure 4.5. The ratio in which glazed to paneled parts present in the door frame is generally kept 2:1 i.e. one- third part of the door has paneled while two-third part consists of glazed.

The paneled part is kept at the bottom half of the door frame and the glazed part is placed on the top side of the door frame. Groove is provided in the shutter for placing glazed parts and these are further secured by placing beading or rails putty. To know more, scan the code.

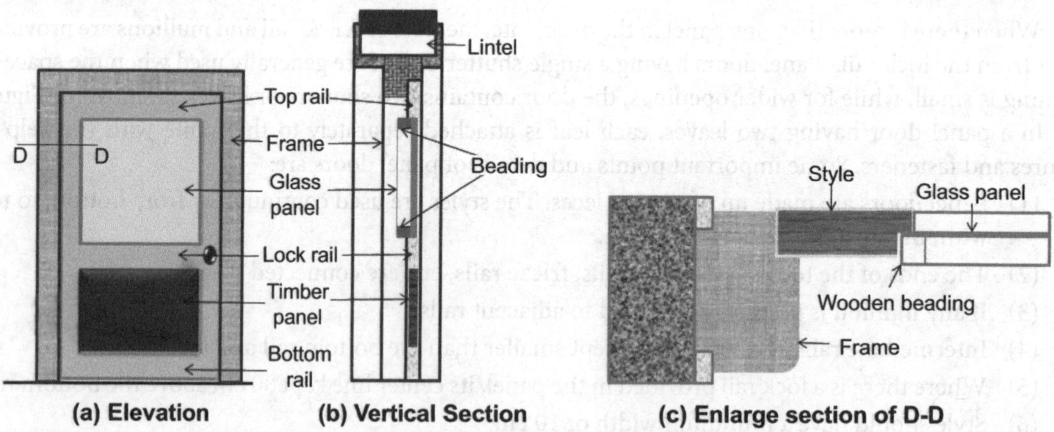

(a) Elevation **(b) Vertical Section** **(c) Enlarge section of D-D**

Fig. 4.5: Details of the partly glazed and partly paneled door

4.6 COLLAPSIBLE DOORS

The collapsible doors are made of mild steel or cast iron. This door is moved horizontally by slight push or pull for closing and opening. These are used to provide extra protection and safety to the godowns, workshops, public buildings, etc. No hinges are required by this type of door for opening and closing as well as no frame is required. The collapsible doors can also be used in a residential building where the opening is wide and framed doors with two shutters cannot be provided. This type of door can be made of a single shutter or double shutter. Figure 4.6 shows the double shutter collapsible door. Vertical double channels of size 20 mm×10 mm×5 mm are used to construct this type of door. They are joined together using iron flats, which are connected to the channels diagonally. The size of the iron flats is 10 to 20 mm with 5 mm thickness. The horizontal distance between the vertical bars is kept 10 cm to 12 cm. There is no restriction for the width of the collapsible door but the height is kept at a maximum of 3 m. T-shaped iron rails are used in which the shutter moves with the help of rollers. These iron rails are fixed at the top and bottom of the door. Handles and locking arrangements are provided in vertical channels.

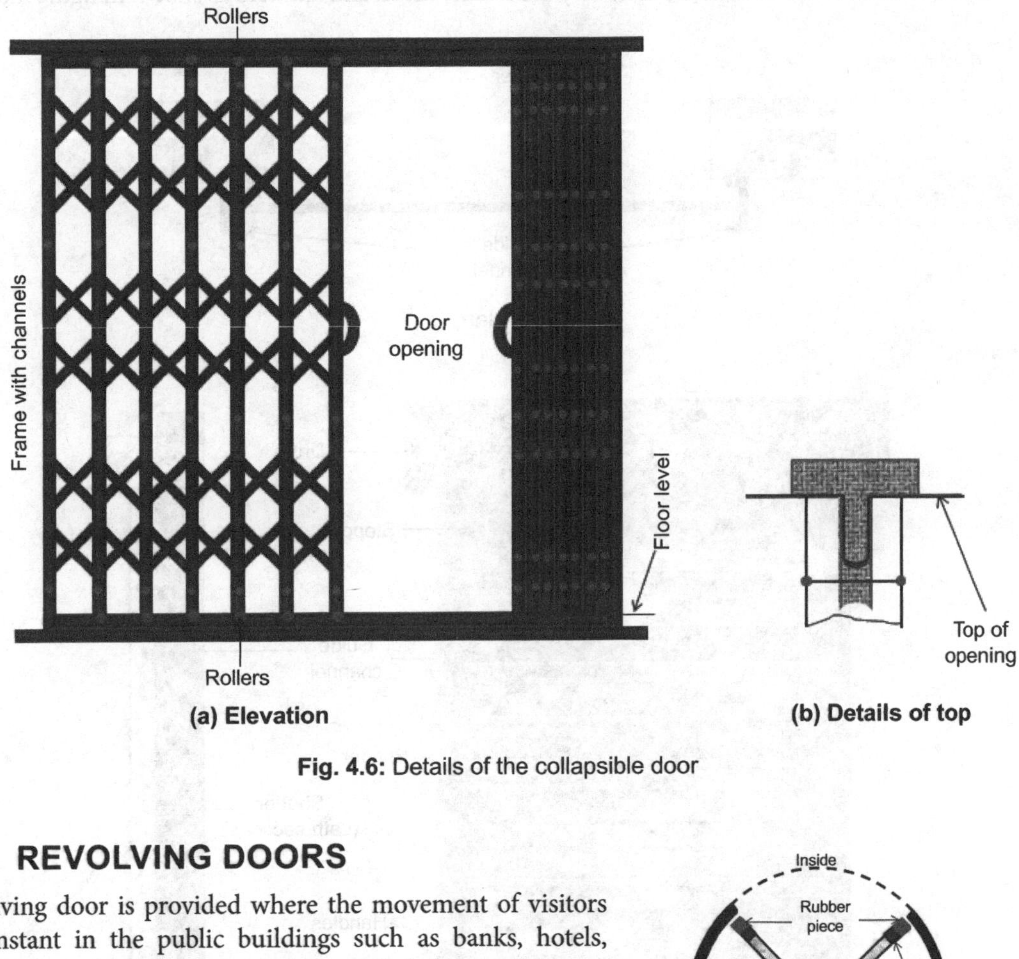

Fig. 4.6: Details of the collapsible door

4.7 REVOLVING DOORS

Revolving door is provided where the movement of visitors is constant in the public buildings such as banks, hotels, museums etc. These doors simultaneously help one visitor to enter and other visitor to exit from the buildings. These doors are generally found in the air-conditioning building or the building located in the place where there is strong wind occurs throughout the year. This door consists of four shutter or leaf attached radially to the mullion placed in the centre as shown in figure 4.7.

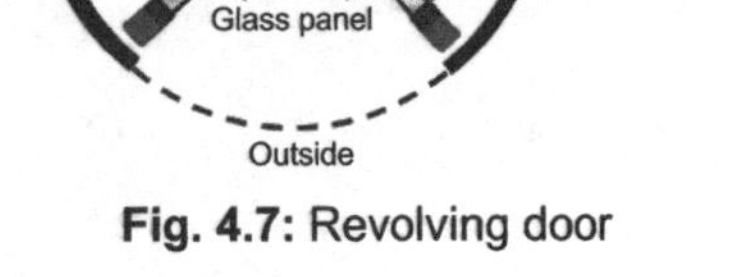

Fig. 4.7: Revolving door

The top portion of the door contains bush bearing while bottom side have ball bearing for supporting the rotation of centrally placed vertical member or mullion without any friction or noise. Shutter provided in revolving door can be fully glazed or partially paneled and partially glazed. Rubbing end of the shutter have rubber at the end to prevent air drought condition.

4.8 ROLLING SHUTTER DOORS

Rolling shutter types of door are commonly used where opening size is very large such as factories, warehouses, shops etc. These types of doors is very strong and provide complete safety against theft and fire. Rolling shutter consist of three parts, they are frame, drum and shutters as shown in figure 4.8.

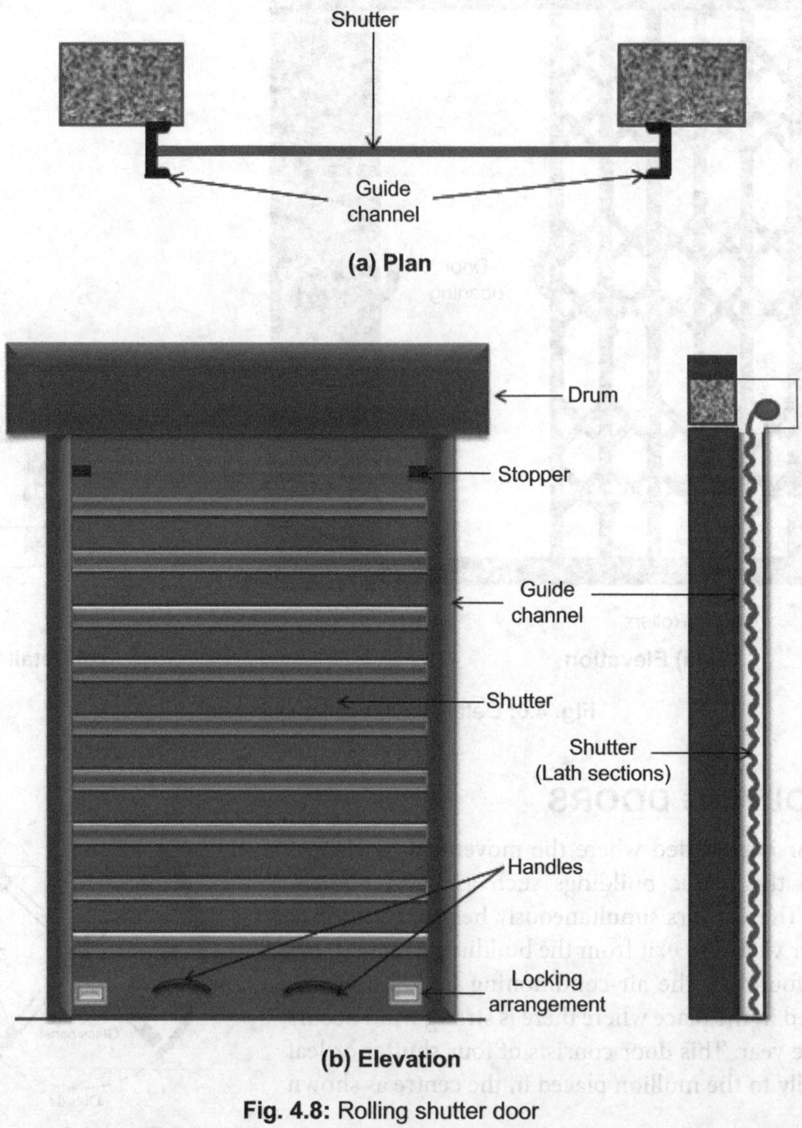

Fig. 4.8: Rolling shutter door

Shutters are made up of thin steel plates known as laths, having a thickness of around 1 mm to 1.25 mm. These steel plates are interconnected and hinged for maximum strength. The frame of rolling shutter contains guided steel channels on both sides in which the shutter moves. Drum consists of springs and shafts which help in shutter movement by push and pull. Drum diameter varies from 200 mm to 300 mm. Rolling shutter based on the operation of shutter is of two types:

- **Push-pull type:** This type of operation is provided when the opening size of the door is not more than 10 square meters.
- **Mechanical gear shutter:** This type of doors is provided for large area. Winding handles, connecting rod, worm gears or chain pulleys system are used in operating this shutter. Very heavy rolling shutters (25 kg/m^2 to 30 kg/m^2) used this mechanism.

4.9 FLUSH DOORS

Flush doors are commonly used in residential, commercial and public buildings as they have smooth surfaces which give attractive appearances and construction is simple, cost-effective and provides better durability. Flush doors are made up of plywood or face veneers at outer surface and inner core is filled with solid or semisolid cardboard or hardwood. The advantages of flush doors are enlisted as follows:

- Cost of construction is less as compared to other type of doors and easily available.
- They are rigid, have high strength and resistant to impact.
- Weight of door is less as compared to plywood door and easily installed.
- These doors can be used as exterior as well as interior door.
- Attractive in appearance and decay-proof

The disadvantages of flush doors are enlisted as follows:

- These doors are difficult in repair due to defects caused by moisture, temperature etc.
- These doors are manufactured in factory, so available in standard sizes and difficult in alteration.
- These doors are not used where door are exposed to rain and sun directly.

The flush doors are of the following two types:

(a) Solid/laminated core flush door

(b) Hollow core/cellular core flush door

(a) Solid/laminated core flush door

Solid/laminated core flush doors are shown in figure no. 4.9. Styles and rails of door frame is made up of hard wood. Inner core of door is made up by glued timber strips under pressure and timber faced is attached to plywood sheets, black board or combination of both plywood and board. Inner core of laminated flush door have wooden strips of 25 mm width and length equal to the laminated core are glued under great pressure. Each core is covered using cross bands and face veneers attached with the help of glue to the both faces.

Generally the grains of cross bands are kept perpendicular to core, while grains of face veneer is kept perpendicular to the cross band. The width of styles, top rails and bottom rails in these type of door is kept a minimum of 75 mm. Solid flush doors are heavy in weight and having high strength. Due to these properties they are used as exterior doors in buildings. They are sound proof and water proof. These doors require less maintenance and more durable.

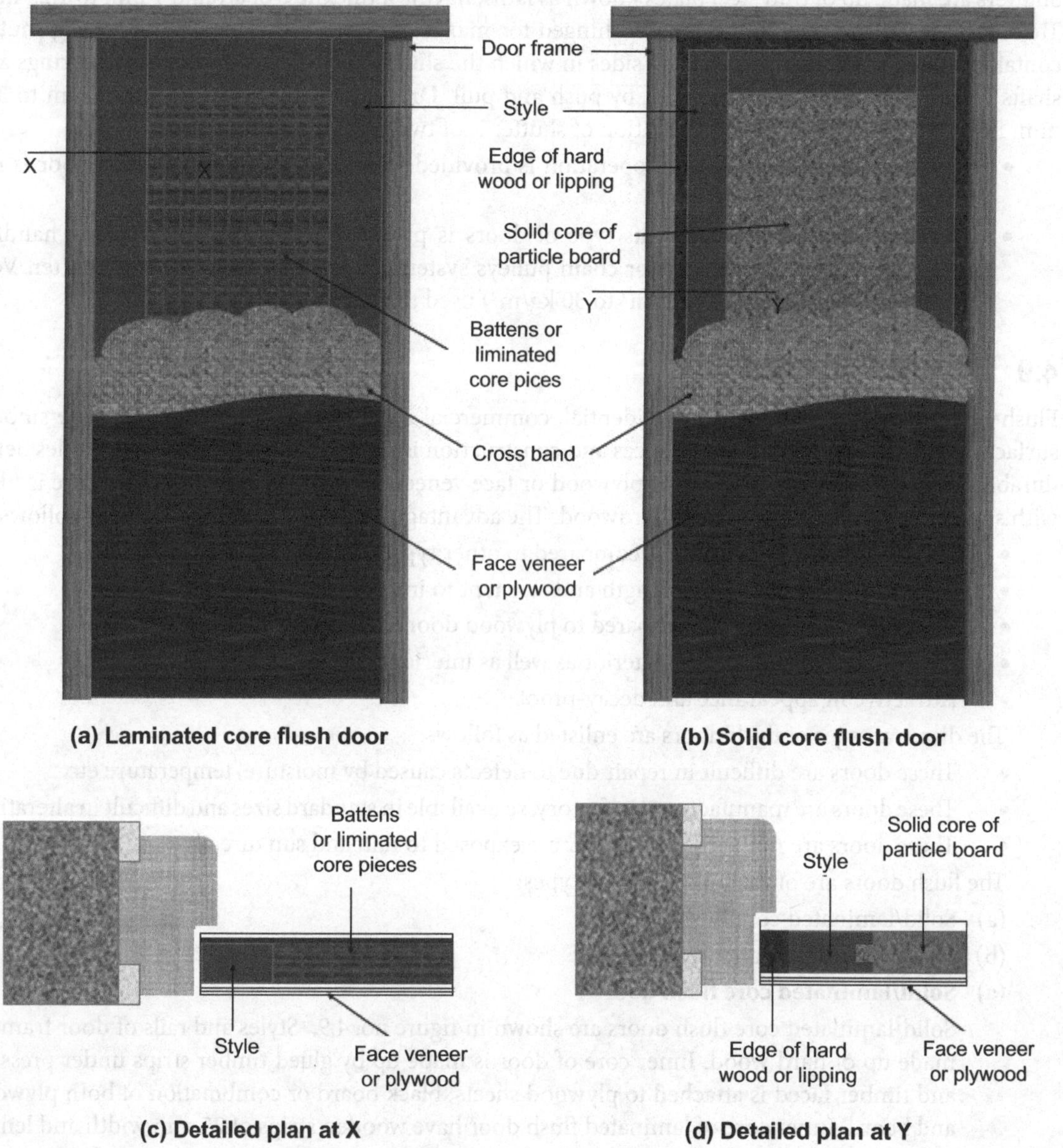

Fig. 4.9: Solid/Laminated core flush door

(b) Hollow core/cellular core flush door

Hollow core flush doors also contain styles, top and bottom rails as used in solid core flush door. This type of door also contains minimum of two intermediate rails having a width minimum of 75 mm and wooden battens of size 25 mm in width equally spaced. The wooden battens and intermediate rails are paced in between the top and bottom rails, so that the area of voids limited to 500 cm^2.

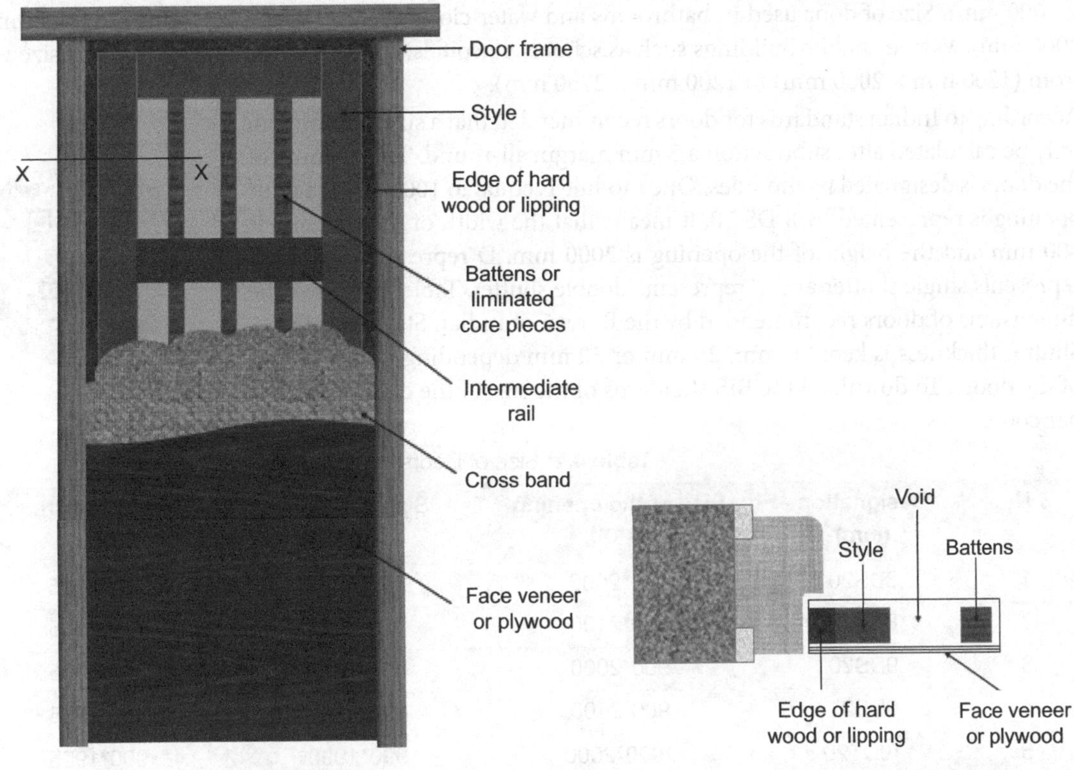

(a) Elevation

(b) Detailed plan at X

Fig. 4.10: Hollow core flush door

Plywood used in this type of door should have a thickness of 6 mm. Hollow core flush door are light in weight than solid core flush door. Cellular core flush doors contain plywood battens of size not less than 25 mm. These battens are placed in hollow part of door such that the area of voids between horizontal and vertical plywood batten not more than 25 cm^2 and the total area of voids should not exceed that 40% of total area of the door. To know more about the hollow core flush door, scan the bar code.

These types of doors are light in weight and provide traditional appearance. In hollow core and cellular core flush door, plywood wood sheets, cross band and face veneer are glued to the both faces of the core under pressure. Figure 4.10 shows the hollow core/cellular core flush door.

4.10 SIZE OF DOORS (AS PER BIS NORMS)

The sizes of the doors are standardized by Bureau of Indian Standards (BIS). The size of the doors in the building is such that the movement of two persons can occur simultaneously without any hindrance. However, the sizes of the door varies for the different types of buildings. The dimensions of the doors must fulfil the purpose of the door. The common ratio for width to height adopted in India for doors is width/height equal to 0.4 to 0.6. General size adopted for external doors in residential buildings in India is (1000 mm × 2000 mm) to (1100 mm × 2000 mm), internal door (900 mm × 2000 mm) to (1000 mm × 2000 mm). Size of door used in bathrooms and water closets is (700 mm × 2000 mm) to (800 mm × 2000 mm). Various public buildings such as schools, hospitals, colleges, libraries, etc. having a size vary from (1200 mm × 2000 mm) to 1200 mm × 2250 mm).

According to Indian standards for doors recommended that a size of doors can only be calculated after subtraction a 5 mm margin all-round. The opening of the doors is designated by modules. One module is equal to 100 mm. If a door opening is represented by 8 DS 20, it means that the width of the opening is 800 mm and the height of the opening is 2000 mm. D represents a door, S represents single shutter and T represents double shutter. Table 4.1 shows the dimensions of doors recommended by the Bureau of Indian Standards (BIS). Shutter thickness is kept 20 mm, 25 mm or 30 mm depending upon the size of the doors. To download the BIS standards of the size of the doors, scan the bar code.

Table 4.1: Size of Doors

S.No	Designation (mm)	Size of the opening (mm)	Size of frame (mm)	Size of shutter (mm)
1	8DS20	800*2000	790*1990	700*1905
2	8DS21	800*2100	790*2090	700*2005
3	9DS20	900*2000	890*1990	800*1905
4	9DS21	900*2100	890*2090	800*2005
5	10DT20	1000*2000	990*1990	900*1905
6	10DT21	1000*2100	990*2090	900*2005
7	12DT20	1200*2000	1190*1990	#1000*1905
8	12DT21	1200*2100	1190*2090	#1000*2005

500 mm for each shutter and 20 mm when closed

4.11 WINDOWS

A window may be defined as an opening in the wall that is made for the entry of natural light and fresh air into the building. A window also provides an outside view to the residents of the building. The windows also make the buildings aesthetically pleasing.

The only disadvantage of the windows is that a window can lose 12 times more energy than a wall. The following points should be kept in mind before deciding the location and type of windows:

- The glass is used in the windows so that the light can enter even if the windows are closed.
- The windows should be provided at the parallel wall for cross ventilation of the air.
- The height of the sill of the window in the rooms is kept 60 cm to 90 cm above the floor. It helps the residents to get a view from the outside.
- The height of the windows in the bathroom is kept around 2 m above the floor.
- The grills are provided at the windows for safety purposes.
- The sun shade can be provided to prevent the entry of rainwater into the rooms.

4.12 COMPONENT OF WINDOWS

The windows have a number of components. However, every window has two major parts which are the frame and shutter. A frame is fixed in the wall and a shutter is the movable part of the window which consists of glass and all the other components. The different components of a typical window can be seen in figure 4.11 and are explained as follows:

(1) **Architrave:** Architrave is a decorative surface that is provided around the opening of the window. The architrave is a well-designed decorative part of the window and can be provided for giving the window an aesthetically pleasing look. However, it is not essential to provide an architrave.

(2) **Reveal:** It is the area provided around the window. It can be cut into the wall. The reveal is the area that remains perpendicular to the window. If the walls have more depth, the architects will get more area for making the reveal.

Reveals are also aesthetically pleasing and can be used for many purposes. These can be provided in the interior and exterior of the walls.

(3) **Frame:** The frame is fixed in the wall while the shutter moves only. It is also known as chaukhat.

(4) **Jambs:** The jamb is the internal part of the window. The sides of the window frames which are vertical are called jambs.

(5) **Head:** The top of the window frame representing the highest point in the window is known as the head. The head remains parallel to the window sill. The height of the window is measured by measuring the distance between the head and the sill.

(6) **Sill:** A sill is the base of the window frame. The sill drains the water during rainfall. The sill can be used to put small things or vases for decoration. If the sill is extended, a reveal is formed.

(7) **Apron:** The apron is constructed to support the sill. It remains just below the sill.

(8) **Casement:** A casement is the section of the window. The top and bottom of the casement are made of rails and the vertical sides are made of stiles. It can be opened at the top or bottom or on the sides.

Fig. 4.11: Components of the window

(9) **Pane:** A pane is the glass installed in the window. It is the glass area of the window. Panes are the central part of the window and remain transparent in nature generally.

(10) **Mullion:** Mullion separates the casements with a vertical partition. Mullions can be made of the same material as that of the casements or different materials can be used. If the number of casements in a window is more than one, these are separated vertically with mullion.

(11) **Muntin:** Muntins are also known as glazing bars which are thin parts of the window. These separate small individual panes.

(12) **Sash:** If the two sections slide one over the other, it is called a window sash. The sashes hold the window pane and the widows can slide over each other in the vertical direction.

(13) **Transom:** The transom divides the window part horizontally. It helps in opening of the upper window independently of the lower window.

4.13 TYPES OF WINDOWS

The frame of the window remain fixed in the wall while the window can move in different ways. The movement of shutter of the window can be used for classification of the windows. Fixed window remains fixed with the frame and can't be opened. Glass is fixed in the window and the fixed windows fulfil the purpose of entry of light in the room. The shutter of the outside opening window opens in the outer direction while the inside opening windows open inside the room. Windows can be hinged at the top or bottom. Sometimes, the shutter of window can be moved with help of roller or slide strips. Such type of window is known as a sliding window. The different types of windows that are used generally are explained in the upcoming sub-sections.

4.13.1 Full Paneled Window

The full paneled window is the oldest form of window which is still in use throughout the world. The full paneled window is made of wood. Currently, polyvinyl chloride (PVC), plywood, aluminium, etc. are also used for making the full paneled window. The number of panels in this type of window can be one, two, three, or six. Figure 4.12 shows the full paneled window. The full paneled window offers the following advantages:

- The strength of the full-paneled window is high. It is strong comparatively.
- It blocks the view of the outsider completely. So, it is safe from the point of view of security.
- The timber is easily available. The different types of timbers for making this type of window can be selected based on the budget of the user.
- The timber can be cut into any desired shape.

There are some limitations of the full paneled windows. The limitations of the full paneled window are enlisted as follows:

- The full paneled window blocks the light completely if it is to be closed.
- The weight of this type of window is very high due to the solid wood panels.
- The material for full paneled window is costly compared to glass.
- There is a risk of insect attack on the wood.

Fig. 4.12: Full paneled window

4.13.2 Glazed Window

In glazed window, glass is used in the panels. The glazed windows are widely used and provide the facility of sunlight entry even if it is closed. It is aesthetically appealing and widely used in the residential buildings, hotels and government buildings. The various types of glass like frosted glass, tinted glass, etc. are used in the glazed windows. The frame of the window can be made of aluminium or timber and glass panels are installed in the frame. Figure 4.13 shows the glazed window.

Fig. 4.13: Glazed window

The glazed windows have a number of advantages as compared to the full paneled window. The limitations of the paneled windows are removed in the glazed windows. A glazed window offers the following advantages:

- It is light in the weight. So, it can be used for sliding window too.
- The light can enter in the glazed windows even if it is closed. It provides the facility of natural lighting of the room.
- The residents can get the outside view even if the window is closed. It prevents the entry of pollution, pollen, rain, etc. but still gives light and the view of outside.
- It is less secured but the security of glazed windows can be increased by providing double glass coating and grills can be provided in front of the glass.
- Glazed window provides more stylish and aesthetically pleasing look.

With the number of advantages, the glazed windows have some limitations too. The limitations of the glazed windows are enlisted as follows:

- The glass windows are not as solid as paneled windows. These offer lesser strength and security.
- There is requirement of curtains for privacy in the rooms due to its transparent characteristics. However, the different types of glass can be used for blocking the view from outside.
- The flexural strength of the glazed window is less than the paneled window.
- The glass can resist the scratches efficiently and it is also easy to clean.

4.13.3 Partly Paneled Window

The partly paneled window is a hybrid form of the paneled and glazed window shown in figure 4.14. The lower portion of the partly paneled window is made of timber while the upper part is made of glass. So, the light can enter the room from the top of the window. Privacy is maintained by the lower part of the window however the upper part gives the advantage of a glazed window.

The partly paneled is moderately weighted and lighter than the full-paneled window but its weight is more than the full-glazed window. It can prevent rainfall, but allows the entry of light even if it is closed. It is more secure than the full-glazed window but less secure than the full-paneled window. It is aesthetically pleasing and stylish. It is more economical than the full-paneled window.

Fig. 4.14: Partly paneled window

4.13.4 Wooden Windows

The windows and their frames can be made of a number of materials. The majorly used material for making the windows are wood, steel, and aluminum. Every material has its own advantages and limitations. As discussed in the previous section, wooden sections can be used for making the frames of the window and the panels of the windows too. Wood as a material for making windows has the following salient features:

- Timber is a universal material and it is available easily.
- Timber is a good insulator of heat. It has low thermal conductivity.
- Timber windows are durable and strong. They have an age of 20 years to 30 years if made from well-seasoned wood.
- Timber acts as a good sound insulator. So, it reduces the noise from outside the building.
- The surface finishing using paints and varnishes is required for the wooden windows. So, the wooden windows can be painted in any desired colour.
- There are chances of insect attack on the wood. The termite attack is common in the case of timber windows.
- The timber is moisture sensitive and it can change its dimensions due to moisture. The wooden windows can bend or wrap in damp conditions.
- The timber requires proper maintenance for long-lasting.

4.13.5 Steel Windows

Steel is a highly durable material for making windows. The steel is manufactured in controlled conditions in the factories. The steel windows have the following salient features:

- The strength of the steel is very high. It can support large glass panels.
- Due to its high strength, steel can be provided in thin frames. It maximizes the viewing capacity of the window.

- The weight of steel windows is high as compared to the wooden frames.
- The steel windows are not prone to the attack of insects.
- It is not a good insulator to heat. So, it is not as energy efficient as wood.
- The cutting of the steel is difficult. The work on site is difficult in the case of steel.
- There is a possibility of corrosion in steel windows. So, it should be treated for making it corrosion-proof.

4.13.6 Aluminium Windows

Aluminium is widely used as a window material nowadays. Due to its lightweight and flexibility, it is a good alternative to wooden and steel frames. The aluminium windows have the following salient features:

- Aluminium is an economical material as it costs less than wood.
- The light weight of the aluminium frames is their most considerable advantage. Due to its lightweight, it can be transported and handled easily.
- The aluminium can be cut in any desired shape. The cutting operations in the case of aluminium are very easy as compared to steel.
- Due to their lightweight, the aluminium windows are easy to install.
- The aluminium windows are durable and strong. So, these can be made in thin frames which maximize the viewing capacity of the window.
- It is also not a good insulator to heat like steel. So, it is not as energy efficient as wood.
- Aluminium is susceptible to corrosion. So, treatment for the prevention of corrosion is required for aluminium windows.

4.13.7 Sliding Windows

The sliding windows consist of shutters that can move horizontally or vertically with the help of roller bearings. Figure 4.15 shows the sliding window.

Fig. 4.15: Sliding window

The sliding windows have the following salient features:

- A sliding window is very easy to operate. Senior citizens and physically handicapped people can use sliding windows as these require less force to open and close due to the roller-bearing mechanism.
- A sliding window is provided with high-quality glazing that makes it highly energy efficient. It can act as a barrier to sound and a good insulator of heat.
- Aluminium is used in sliding windows generally. The sliding window frames have the ability to hold large glass sections, which maximizes their viewing capacity.
- Due to the large glass area, the sliding windows provide a good amount of natural light in the room.
- The sliding windows require less maintenance comparatively. However, the durability of the window depends on the material used in the window frame and the quality of the glass.

4.13.8 Louvered Windows

The louvered windows have louvers or blades that can be adjusted at any equal angle. When it is closed it gives an equal amount of air throughout the room and privacy is maintained. The blades in the louvered windows can be fixed or adjustable. In the case of fixed blades, these are kept at an angle of 45°. The inclined blades help in draining the rainwater too. The louvered windows can be used in toilets, bathrooms, and factories. Figure 4.16 shows the louvered window. The louvered windows have the following salient features:

- The louvered window provides the facility for the passage of air and light uniformly. The flow of air is maintained in the room even if it is closed.
- The flow of the air can be controlled in the building if the blades of the window are movable in nature.
- The louvered windows are easy to clean. These can be cleaned from the inside of the room.
- This type of window can customize the privacy of the residents. The louvers can be inclined at any desired angle.

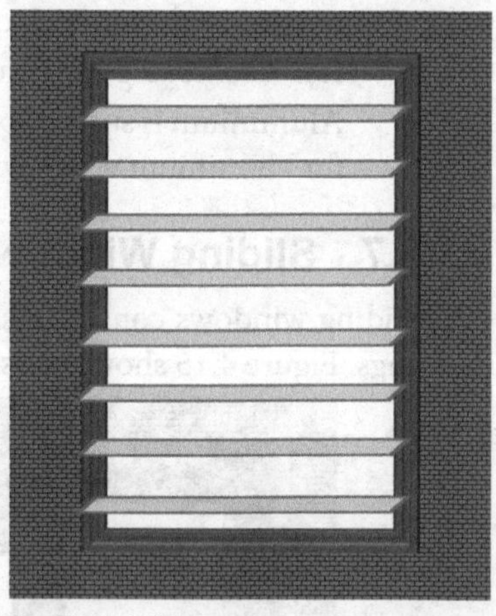

Fig. 4.16: Louvered window

4.13.9 Bay Window

A bay window is projected outside the building at any angle. The bay window is a combination of windows that extends beyond the external wall in triangular, rectangular or polygon form. Figure 4.17 shows the bay window. A bay window has the following salient features:

- A bay window is more stylish comparatively. It is aesthetically pleasing.
- The bay window has an enlarged area outside the wall. So, the room gets a sufficient amount of natural light.
- The bay window provides additional space that can be utilized for different purposes.

Fig. 4.17: Bay window

- The bay window provides good ventilation to the building. As it is a combination of windows, it can provide good airflow.

4.13.10 Corner Window

The corner window is a combination of two windows that are provided at the corner of the building and remains perpendicular to each other. This type of window can be provided in small or large sizes depending on the size of the structure. The faces of the windows meet perpendicularly. Figure 4.18 shows the corner window. A corner window has the following salient features:

- The corner windows provide uniform light and ventilation throughout the room.
- The corner windows are aesthetically pleasing and appealing from the architectural point of view.
- These types of windows provide natural light to the interior places of the house.
- The corner windows are easy to install.
- The corner windows require less maintenance as the amount of dust remains less in the corners.
- The corner windows are suitable for the kitchen, dining hall, and any part of the house that needs daylight.

Fig. 4.18: Corner window

4.13.11 Clerestory Window

Clerestory window is pronounced as clear story window. The clerestory window is provided near the ceiling. The major function of the clerestory window is to remove the contaminated air from the room. The window is pivoted both the sides and can be opened from the bottom. Clerestory window is a ventilator. Figure 4.19 shows the clerestory window. This type of window has the following salient features:

- The clerestory window is provided for removing the contaminated air. So, it provides a better circulation of air.
- This type of window provide sufficient amount of natural light in the room.
- It is highly energy efficient.
- As it is provided near the ceiling, it provides privacy to the residents.
- Due to the window's height, the heating problem due to sunlight can be more in case of clerestory window.
- The design of the clerestory window is costly. The wall slab junction becomes critical due to the installation of a clerestory window.

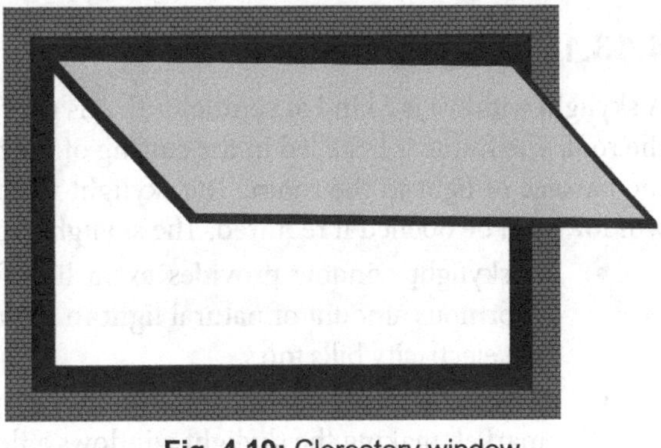

Fig. 4.19: Clerestory window

4.13.12 Gable and Dormer Window

A gable window is provided at the triangular part of the sloping roof. It has a flat surface. The gable windows provide sufficient natural light to the room. The dormer windows are projected in nature and are provided at the sides of sloping roofs. The gable and dormer windows have the following salient features:

- The gable and dormer windows are aesthetically pleasing. These increase the beauty of the home architecturally.
- The gable windows are easy to install and remove.
- The dormer windows are resistant to rain and snow. These can easily drain off the rainwater and snow due to their slope.
- The cost of construction of gable and dormer windows is not very costly. These types of windows are economical. Figure 4.20 (a) and (b) show the gable and dormer windows respectively.

Fig. 4.20 (a): Gable window

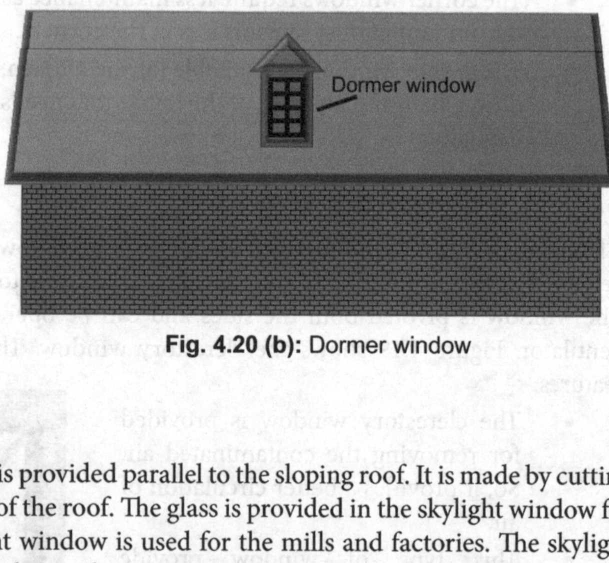

Fig. 4.20 (b): Dormer window

4.13.13 Skylight Window

A skylight window is a kind of ventilator that is provided parallel to the sloping roof. It is made by cutting the roof. The frame is installed in the cutting of the roof. The glass is provided in the skylight window for the passage of light to the room. The skylight window is used for the mills and factories. The skylight windows can be opened if required. The skylight windows offer the following salient features:

- A skylight window provides extra light to the room. Due to the enormous amount of natural light in the room, it helps in reducing the electricity bills too.
- It is aesthetically pleasing. Many design options are available in the market, making the skylight windows a flexible option.
- The skylight windows can overheat the rooms due to the excessive amount of sunlight. To know more about the skylight window the bar code can be scanned.

SCAN ME

4.14 SIZES OF WINDOWS (AS PER BIS NORMS)

The area of the windows depends on the size of the room. The total area of the windows for a residential building should be at least 1/8 of the floor area while the area of glazing should be at least 1/10 of the floor area. The area of the ventilator is kept 1/25 of the area of the floor of the room. The area of windows in bathrooms and toilets is kept 1/10 of the area of the floor. The minimum area of the windows for the factories is kept 15% to 25% of the floor area of the room while for the hospitals it should not be less than 20% of the area of the room. The opening of the windows is designated by modules. One module is equal to 100 mm. If a window opening is represented by 6WS12, it means that the width of the opening is 600 mm and the height of the opening is 1200 mm. W represents a window, S represents single shutter and T represents double shutter. Table 4.1 shows the dimensions of windows recommended by the Bureau of Indian Standards (BIS).

Table 4.2: Size of Windows

S.No	Designation (mm)	Size of the opening (mm)	Size of frame (mm)	Size of shutter (mm)
1	6WS12	600*1200	590*1190	500*1100
2	10WT12	1000*1200	990*1190	460*1100
3	12WT12	1200*1200	1190*1190	560*1100
4	6WS13	600*1300	590*1290	500*1200
5	10WT13	1000*1300	990*1290	460*1200
6	12WT13	1200*1300	1190*1290	560*1200

4.15 VENTILATORS

An adult breathes at least 16 times in a minute. The air in the buildings can get polluted due to the CO_2 produced by humans, outdoor pollution can enter the house and the equipment like furniture. In a home, the residents require clean air. In industrial buildings, indoor pollutants are to be excluded from the building. So, ventilators are provided to remove the contaminated air from the building. The ventilators are the same as small windows but they are provided about 30 cm to 50 cm below the ceiling. The ventilation in the building can be provided using the following techniques:

(1) **Natural ventilation:** A small window is provided below the ceiling for the purpose of natural ventilation. The difference between the pressure due to wind and the temperature difference due to the inside and outside of the building helps in creating natural ventilation. Natural ventilation can offer a sufficient flow of air throughout the building. However, the natural ventilation is subject to change due to external atmospheric conditions.

(2) **Mechanical ventilation:** As we know that natural ventilation depends on the outside condition of the atmosphere. So, mechanical ventilators can be provided to extract the contaminated air from the inside of the building. The pressure difference is created by a dynamic fan in the case of mechanical ventilation. The exhaust fans in the bathrooms and toilets are examples of mechanical ventilators.

(3) **Hybrid ventilation:** It is evident that natural ventilators save energy but depend on atmospheric conditions. So, a combined system with natural and mechanical ventilation is provided which is known as a hybrid ventilation system. A window with the exhaust fan can be provided in the case of hybrid ventilation.

4.16 FIXURES AND FASTENINGS FOR THE DOORS AND WINDOWS

The different fixtures and fastenings are used during the installation of windows and doors. The fixtures and fastenings help in the movement of the doors and windows. The description of various fixtures and fastenings is as follows:

(1) **Hinges:** The hinges are used for fixing the frames with the shutters for the windows and the doors. The hinges are made of metals like aluminium, steel, iron, etc. The different types of hinges are shown in figure 4.21.

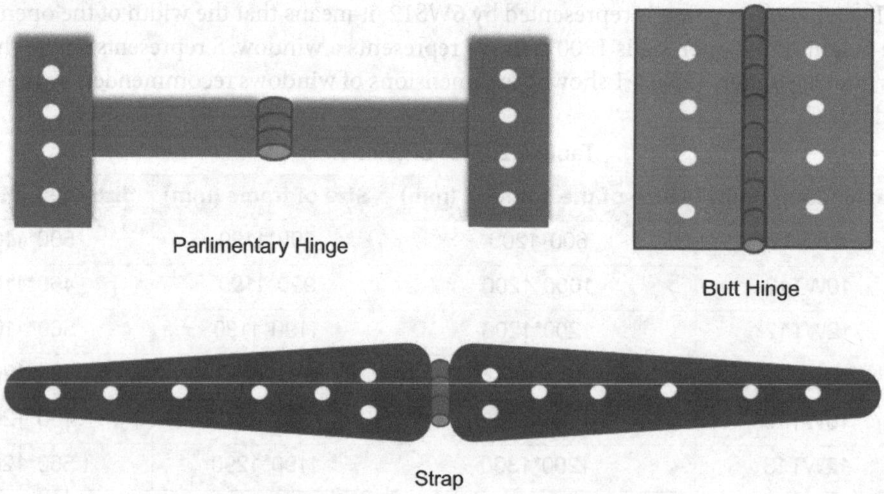

Fig. 4.21: Hinges for doors and windows

(2) **Handles:** The handles are fixed to the shutters. The handles are provided on the outside and inside of the doors and windows. These help in opening and closing of the doors and windows conveniently. Handles can be provided in different shapes and sizes. The stylish and aesthetically pleasing handles increase the beauty of the door and windows. The size of the handles should be sufficient to hold it. There should not be any sharp edge on the handle for safety purposes. Figure 4.22 shows the handles.

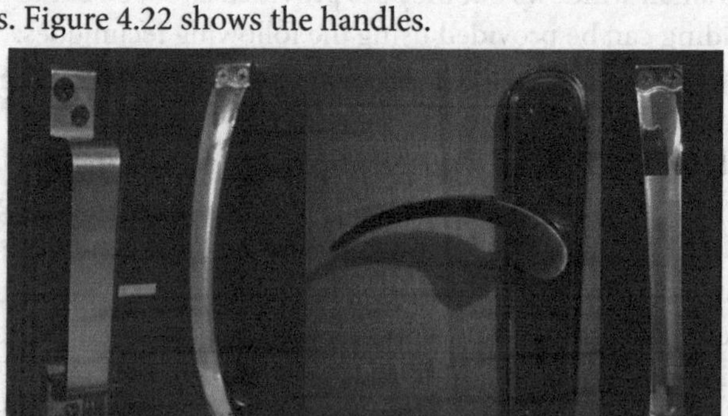

Fig. 4.22: Handles

(3) **Bolts:** The doors and the windows are kept closed using the bolts. There are different types of the bolts that can be used in a door and window. These are made of aluminium, steel or iron. Figure 4.23 shows the different bolts used in the doors and windows.

Barrel bolt

Hasp and staple

Aldrop bolt

Fig. 4.23: Bolts

(4) **Miscellaneous items:** The locks are used to close the door with significant security. There are different types of locks that are used. These provide security and safety to the building. The door stoppers are also provided. The stoppers are used to prevent the sudden closing and opening of the doors under the impact of wind and other forces. The stoppers can be made of wood or rubber as shown in figure 4.24 (a). Figures 4.24 (b) and (c) show the different types of locks.

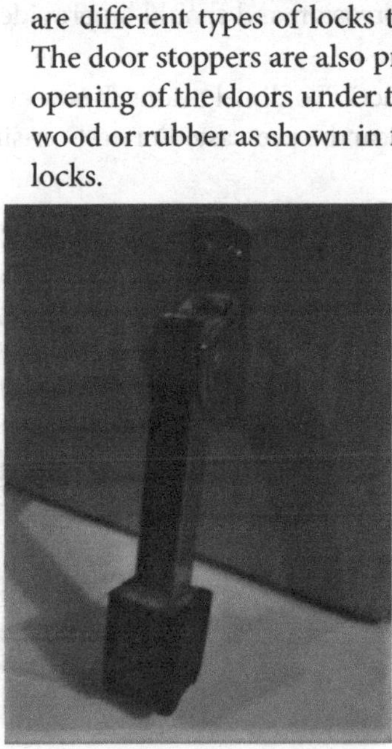

Fig. 4.24 (a): Door stopper

Fig. 4.24 (b): Pad lock

Fig. 4.24 (c): Cup board lock

4.17 FUNCTIONS AND MATERIALS OF WINDOW SILL AND LINTELS

A window sill is constructed at the bottom of the window and is like a shelf under it. The window sills are provided for the following functions:

- The window sill provides a base for the window and its frame. The sill acts as a rigid base which provides stability to the window.
- It provides a slope to drain off the rainwater. The sill helps in the prevention of the entry of water into the room.
- The window sill can be used for different purposes. It can be used for placing decorative pieces.

A window sill is made of reinforced cement concrete however, a stylish and fancy material is provided as a covering to the sill. The granite stone or tiles can be provided on the sill. Stone can be used for providing the sill covering as it provides a smooth surface for placing the decorative on the sill.

A lintel is a beam-like structure that is provided above the window. It takes the load of the bricks above the window. The lintel is provided for the following functions:

- The lintel act as a tie beam that transfers the load of the masonry above the window to the surrounding masonry.
- The lintel acts as a damp-proof course (DPC) and prevents the penetration of water.
- The lintel supports the chajja or sheds.

As the lintels are beams, different materials can be used to construct the lintel beams. Timber is used as lintels since a long time. A bearing of 150 mm to 200 mm is provided for the timber lintels. These types of lintels are prone to fire hazards and can be attacked by insects.

Steel beams are also used as a lintel. The steel beams are light in weight and can be handled easily. The transportation and placement of the steel lintel are easy. It has high tensile strength. The steel can be moulded into any required shape. The steel lintels are prone to corrosion and should be provided with anti-rust coatings.

Nowadays, reinforced cement concrete is used for the construction of lintel beams. It is very strong and can take the tension and compression loads. The steel lintels can be pre-casted or cast on-site. The concrete lintels are durable.

4.18 SHED/CHHAJJA

A chhajja is also known as a sunshade. It is an overhanging eave above the window. It protects the window from the water and the sunlight. It is also called sunblock or eaves. A chhajja has the following functions in a building:

- Chhajja protects the house from sunlight and rainwater. It prevents the entry of rainwater inside the room.
- As the wood is sensitive to moisture, chhajja protects the wooden windows from rainwater.
- It can be designed to give an aesthetically pleasing appearance.
- It can be used for placing the air coolers or air conditioner's compressor unit. To know more about chhajja, scan the bar code.

SCAN ME

4.19 MEANS OF VERTICAL COMMUNICATION

The buildings are constructed with multiple floors nowadays. The buildings require different means for vertical communication. Vertical communication refers to the movement of persons and goods vertically using staircase, ramps, lift, etc. The various modes for the vertical communication are explained as follows:

(1) **Staircase:** Stairs are the essential parts of a building. A stair is a combination of steps that leads from one floor to the other. The overall area that includes the steps, hand rail, landing, etc. is known as staircase. The stairs provide manual access to the different floors of the building. The residents have to use body force to climb the stairs. So, it may not suitable for physically handicapped people.

Fig. 4.25: Ramp

(2) **Ramps:** Ramps are the inclined platforms that are used for transporting the heavy goods, machinery, furniture or physically disabled persons from one floor to the other. The ramp can be provided at the entrance of the building too. It helps in easy entry and exit of the building. The ramps are typically like a floor that is inclined at some angle so that the goods and heavy items can be rolled down on the ramps. The gradient of the ramps in a building is kept from 1 in 10 to 1 in 15. Figure 4.26 shows the ramp. A ramp should have the following characteristics:

- ➢ As discussed above, the slope of the ramps is taken from 1 in 10 to 1 in 15, so the construction of ramps requires large space.
- ➢ The ramps are can be constructed in any geometric shape that is aesthetically pleasing and suitable for the movement of persons and goods.
- ➢ The texture of the ramps must not be very smooth. The smooth surfaces can pose the risk of slipping.
- ➢ The handrails should be provided on the sides of the ramp.

(3) **Lift or elevators:** The stairs have some limitations as humans can climb up to a height of 30 m in one run. However, even after a height of 15 m, the human body feels disturbed. Communication through stairs is also a time-consuming process. Nowadays, due to the shortage of space, multi-storey buildings are constructed in the urban area. So, lifts or elevators are provided in the buildings. It is mandatory to

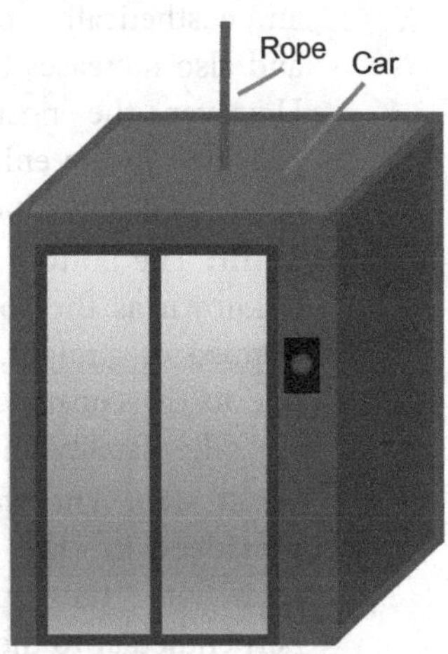

Fig. 4.26: Lift or elevator

provide a lift in buildings of more than 14.5 m in height. The traction-type elevators are used widely. The car is the part of the lift which is used for accommodating goods and people. The car is tied to a thick rope and the rope is attached to a counterweight through a pulley. The electric motor is used to drive the pulley and the lift can be moved up and down. Figure 4.26 shows a lift.

(4) **Escalators:** The escalators are movable stairs. If the number of users per hour is very high, escalators are provided in the building. The escalators are provided in hospitals, shopping malls, business areas, and any place where users are large in number. The user has to stand on the stair while the stair moves and takes the user to the desired floor level. Handrails are provided for safety purposes.

4.20 TERMS USED IN STAIRCASE

A staircase is known as the overall area that includes the steps, handrail, landing, etc. Figure 4.27 shows the details of staircase terms. The different terms that are used in designing and constructing the staircase are discussed as follows:

(1) **Tread:** A tread is the horizontal portion of the stair which is used to climb the stairs. The tread accommodates the foot and hence should be of sufficient width. The typical width of the human foot is 23 cm.

(2) **Riser:** The vertical member between two consecutive treads is known as the riser while the height of the riser is known as the rise.

(3) **Step:** The combination of tread and riser is known as a step.

(4) **Nosing:** It is the extending edge of the tread. The nosing of the tread provides an aesthetically pleasing appearance and also increases the area of the tread. However, the nosing should not be provided in the enlarged form as it can create an obstruction to the foot.

(5) **Soffit:** The bottom surface of the stairs is known as the soffit. A soffit can be stepped or straight. It may be invisible due to the construction of the store or any other facility under the stairs.

(6) **Waist slab:** The waist slab is the slab provided in the stairs. The waist represents the minimum thickness perpendicular to the soffit and the stairs.

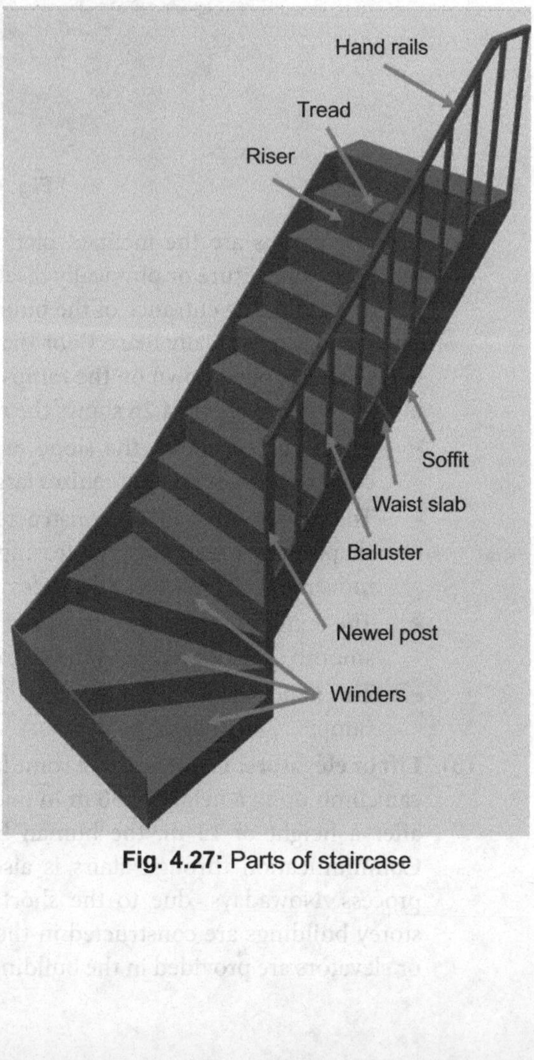

Fig. 4.27: Parts of staircase

(7) **Baluster:** These are the vertical members on the railing. The balusters can be made of various aesthetically pleasing designs. These can be made of wood, concrete, steel, etc.
(8) **Balustrade:** The balustrade is also known as the railing or hand rail of the staircase. The railings are constructed using multiple balusters.
(9) **Scotia:** A wooden block provided under the nosing of the staircase is known as scotia. It imparts strength to the nosing.
(10) **Handrails:** Handrails are provided to support the user of the staircase. It is supported by balusters.
(11) **Newel post:** Newel post is the major vertical member that is constructed for anchoring the handrails. It is provided on the foot, top, and landing of the staircase.
(12) **Landing:** It is a horizontal leveled floor provided at the places where the staircase changes direction. It can be provided at the top, bottom, or at intermediate of the staircase.
(13) **Headroom:** It is the gap between the step and the obstruction at the top. It is also known as head clearance. The headroom should be sufficient to avoid injuries. A headroom of approximately 2 m is sufficient.
(14) **Winder:** When the staircase changes direction, the shape of the steps is modified for taking the turn. This is done by providing the steps with varying shapes known as winders.

4.21 TYPES OF STAIRCASES (ON THE BASIS OF SHAPES)

The staircases are provided in different shapes. The shape of the staircases is selected based on the availability of the space and the aesthetics of the building. The various types of staircases based on the shapes are discussed briefly.

4.21.1 Straight Staircase

The straight staircase consists of steps in a straight line and there is no turn in such type of staircase. The straight staircase provides sufficient visibility to the users and reduces the chances of accidents due to its straightness. The landings are provided so that users can rest if there are a large number of steps. If the wall is not on the side of the straight staircase, the railings are provided. When the available width is less while the length is sufficient for construction, straight staircases can be provided. Figure 4.28 (a) and (b) show the straight staircase respectively.

Fig. 4.28 (a): Straight staircase

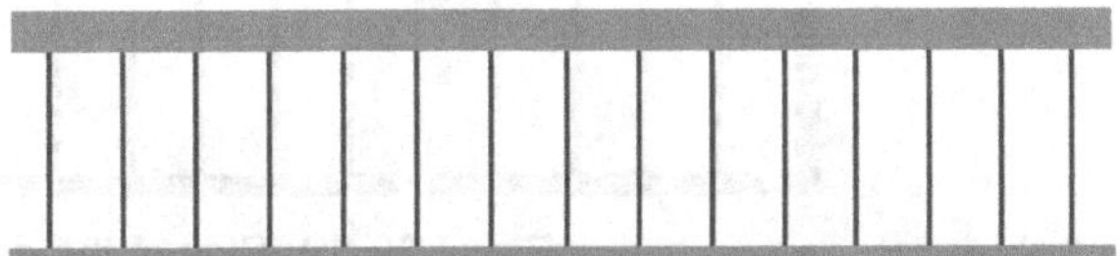

Fig. 4.28 (b): Plan of the straight staircase

4.21.2 Dog-legged Staircase/ Half-turn Staircase

The dog-legged staircase or half-turn staircase consists of two parallel flights connected to a common landing. This type of staircase is also known as a U-shaped staircase. The stairs turn at an angle of 180°. Normally, this type of staircase is constructed in multi-story buildings.

The dog-legged staircase uses the space efficiently and provides privacy as the two floors are separated. These type of staircases require skilled workers. The handrails must be designed carefully for this type of staircase. Figure 4.29 (a) and (b) show the elevation and plan of dog-legged staircase respectively.

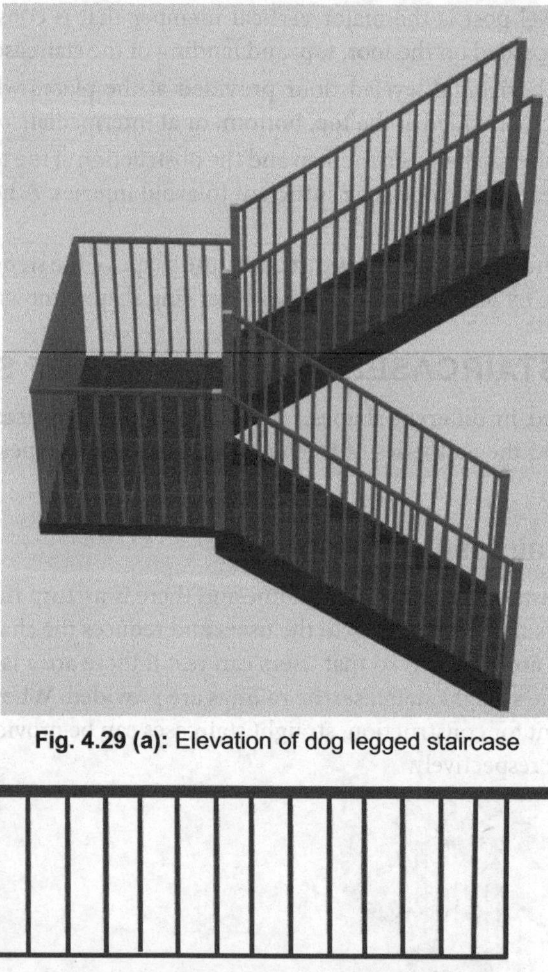

Fig. 4.29 (a): Elevation of dog legged staircase

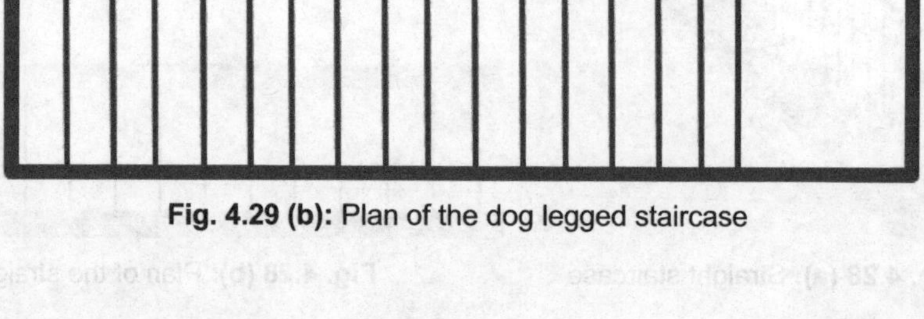

Fig. 4.29 (b): Plan of the dog legged staircase

4.21.3 Open Well Staircase

The open well staircase is the same as the dog-legged staircase but the space provided between two flights is more as compared to the dog-legged staircase. The space between flights looks like a well and the steps are circled around this space. The width of the staircase is twice the width of the step and the width of the well. This type of staircase is provided in public buildings. Figure 4.31 shows the plan of the open well staircase. The open well staircase is also an example of half-turn staircase.

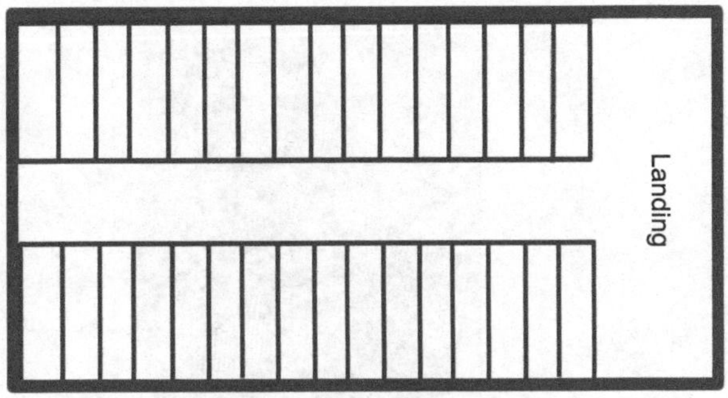

Fig. 4.30: Open well staircase

4.21.4 Spiral Staircase

In the case of spiral staircase, the steps are curved and attached to a newel post. The central shaft is made of iron, steel or wood. The spiral staircase are stylish and aesthetically pleasing. This type of staircase can be customized according to the requirement and available space. The hand railing is provided. If the availability of space is limited, the spiral staircase is provided. The spiral staircase is easy to install and the components of this type of staircase can be constructed by composite materials. The spiral staircase are widely used in the control towers and multi-story mills due to their ability to occupy less space. Figure 4.31 (a) shows the spiral staircase and figure 4.31 (b) shows the plan of the spiral staircase.

Fig. 4.31 (a): Spiral staircase

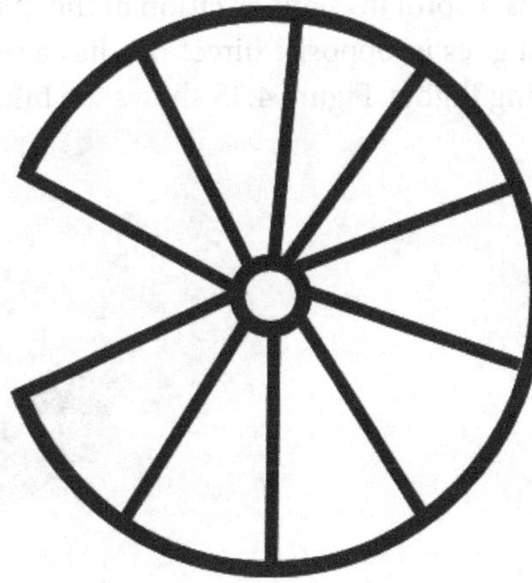

Fig. 4.31 (b): Plan of the spiral staircase

4.21.5 Quarter-turn and Three Quarter-turn

When there is a need of changing the direction of the stairs, the turns are provided at the 90° angle at different levels. In this case, quarter-turn, three-quarter-turn, and, half-turn staircases may be provided.

In a quarter-turn staircase, the stairs are turned through the 90° angle with the help of a level landing. The quarter-turn staircase is used in shops and public buildings.

Fig. 4.32 (a): Quarter-turn Staircase

Fig. 4.32 (b): Three-quarter-turn staircase

In a three-quarter-turn, the stairs turn at 270° by changing the angles multiple times. A large well is formed in the case of a three-quarter-turn staircase. The details of both staircases is shown in figure 4.32 (a) and (b).

4.21.6 Bifurcated Staircase

In the case of a bifurcated staircase, the staircase is constructed in a single line but after some height, it is divided into two opposite directions. In this case, three flights meet at a common landing. The bifurcated staircase is helpful in channelization of the crowd. The width of the bottom flight is kept higher while the flight that goes in opposite directions has a reduced width. It is due to the distribution of crowds at the distributing flights. Figure 4.33 shows the bifurcated staircase.

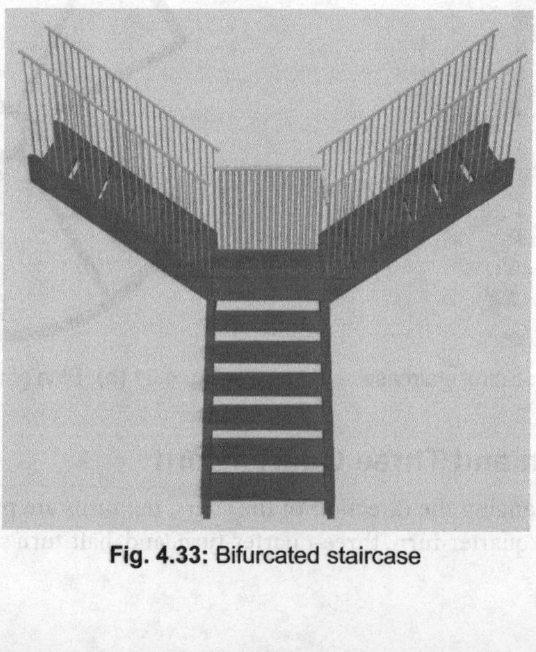

Fig. 4.33: Bifurcated staircase

4.22 TYPES OF STAIRCASES (ON THE BASIS OF MATERIALS)

The staircases are constructed using different materials. The materials like wood, stone, brick masonry, metal and reinforced cement concrete are commonly used in the construction of the staircases. The different types of staircases are discussed in brief.

4.22.1 Stone Staircase

Stone is a durable and strong material that is used for the construction of the staircase. Stone as a material of the staircase offers the following features:

(1) Stone provides stability due to its heavy weight.
(2) Stones are durable and strong in nature.
(3) Stones as a material for the staircase provide an aesthetically pleasing appearance.
(4) Stones offer sufficient resistance to wear and tear. However, the stones for the construction of the staircase must be checked for their wear characteristics.
(5) Stones provide sufficient resistance to fire.
(6) Due to their heavy weight, the stones are difficult to transport and handle.

The simplest form of the stone staircase is constructed by placing rectangular blocks of stone ashlar masonry. The steps are arranged as shown in figure 4.34 (a). The spandrel steps are of near triangular shape and the soffit remains plane. The spandrel steps are aesthetically pleasing. Figure 4.34 (b) shows the spandrel steps. The steps can be made by treads only. There is no riser in this type of staircase. The treads are inserted into the wall from one side and act as a cantilever. Figure 4.34 (c) shows the cantilever staircase. In the built-up steps, the thin stones are placed on the steps of concrete or brick masonry. The thickness of the stone slab may vary from 2 cm to 5 cm as shown in figure 4.35 (d).

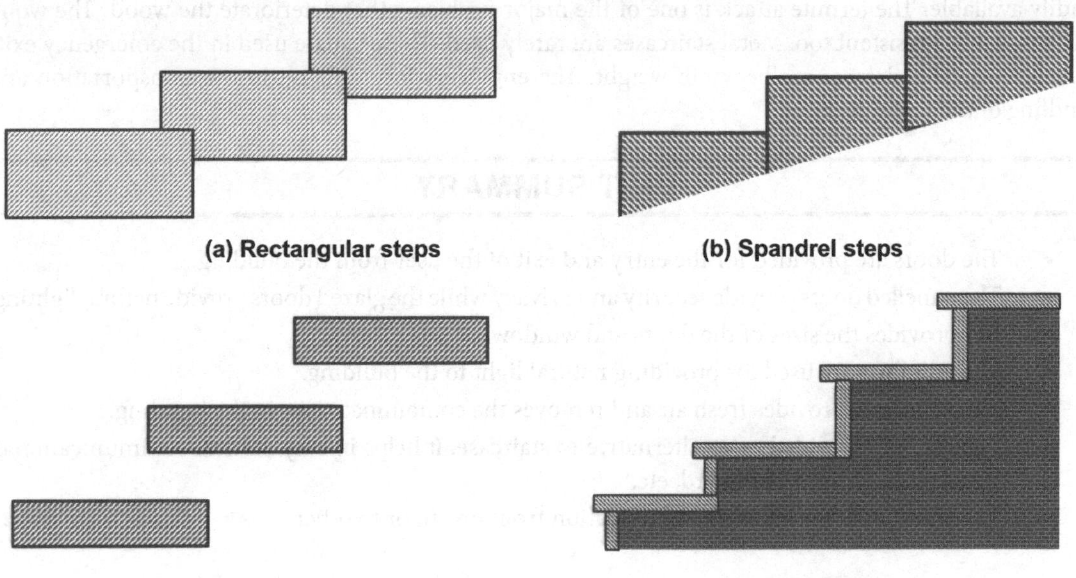

(a) Rectangular steps (b) Spandrel steps

(c) Cantilever steps (d) Built-up steps

Fig. 4.34: Stone staircase

4.22.2 Brick Staircase

Bricks are rarely used as the material for the construction of staircases. The brick staircases as shown in figure 4.35 are used in rural regions. The bricks are brittle and have very poor abrasion characteristics. So, the brick staircases require regular maintenance. The brick staircases are plastered to prevent the wear of the bricks. Stone toppings can also be provided for the brick staircases.

Fig. 4.35: Brick staircase

4.22.3 RCC Staircase

Reinforced cement concrete (RCC) is widely used for the construction of staircases. The RCC steps can be constructed on the site or pre-fabricated. The RCC steps are durable and can take heavy loads. These steps are strong and fire-resistant. The RCC steps can be moulded into any desired shape. These staircases can be constructed as non-slippery surfaces.

The RCC staircases are easy to clean. This type of staircase is aesthetically pleasing. The steps can be provided with stone covering for improving the appearance of RCC staircases.

4.22.4 Wooden and Metal Staircase

Wood is also used widely for the construction of staircases. The wood as a material for the staircase is light in weight and can be transported easily. These types of staircases are aesthetically pleasing and act as home decor. The wooden staircases are economical as the raw material for the wooden staircases is readily available. The termite attack is one of the major problems that deteriorate the wood. The wood is not much fire resistant too. Metal staircases are rarely used. These can be used in the emergency exits only. The metal staircases are heavy in weight. The engineers face difficult in the transportation and handling of the metal steps.

UNIT SUMMARY

- The doors are provided for the entry and exit of the user from the building.
- The panelled doors provide security and privacy while the glazed doors provide natural lighting.
- BIS provides the sizes of the doors and windows.
- The windows are used for providing natural light to the building.
- The ventilator provides fresh air and removes the contaminated air in the building.
- The ramp is provided as an alternative to staircase. It helps in easy vertical communication of loads like furniture, cupboard, etc.
- The staircase are used for communication from one floor to other.

EXERCISES

Multiple Choice Questions

1. Which is the top horizontal member of the shutter?
   (a) Head (b) Jamb
   (c) Top rail (d) Bottom rail
2. The area of the shutter enclosed between two adjacent rails is known as:
   (a) Head (b) Jamb
   (c) Panel (d) Holdfast
3. A vertical member that divides a window opening vertically is known as:
   (a) Head (b) Mullion
   (c) Jamb (d) None of the above
4. The most common type of door is:
   (a) Single leaf door (b) Double leaf door
   (c) Louvered door (d) None of the above
5. The width of the window is selected based on the factor:
   (d) Dimension of room (b) Room location
   (c) Purpose of room (d) All of the above
6. Which type of window shutter opens like a door?
   (a) Gable window (b) Louvered window
   (c) Casement window (d) All of the above
7. What is the minimum height of a door?
   (d) 1.5 m (b) 1.8 m
   (c) 2.0 m (d) None of the above
8. The slab provided above the window and door is called:
   (a) Sill (b) Chajja
   (c) Casement (d) Lintel
9. Sill is provided for:
   (a) Prevention of moisture (b) To give strength
   (c) To reduce wear (d) All of the above
10. A sloping construction provided as an alternative to the staircase is called:
    (a) Lift (b) Tread
    (c) Rise (d) Ramp

Answers of Multiple Choice Questions

1. (c), 2. (c), 3. (b), 4. (a), 5. (d), 6. (c), 7. (b), 8. (d), 9. (a), 10. (d)

Short and Long Answer Type Questions

1. Write a short note on the importance of doors in the building.
2. Write a short note on fully panelled door.
3. Write a short note on fully gazed door and its advantages.
4. Explain the suitability of the revolving door.
5. Explain the flush door with a neat sketch.
6. Write a short note on the louvered window.
7. Enlist the different advantages and disadvantages of timber windows.
8. Discuss the different types of ventilation in a building.
9. Explain the different types of locks with a neat diagram.
10. Draw a neat sketch of a staircase and label its different components.
11. Write a short note on half turn staircase.
12. Explain the advantages of quarter turn staircase.
13. Draw a neat sketch of bifurcated staircase. Also discuss its suitability.
14. Explain the concept of skylight window.
15. Write a short note on escalators.
16. Discuss the vertical communication using lift in the building.
17. Explain the different types of materials that are used for the construction of staircases.
18. Discuss the various functions of the window sill.
19. Explain the concept of lintel. Also discuss the purpose of providing a lintel.
20. Explain the different advantages of RCC staircases.

KNOW MORE

As discussed in this unit, spiral staircases are widely used. These are aesthetically pleasing and stylish. Vatican museums have spiral staircases that were constructed in the 16th century. Do you know about the most famous spiral staircases in the world? To know more about the most famous staircase of the world, scan the QR code given.

SCAN ME

REFERENCES AND SUGGESTED READINGS

1. Chudley, R., Greeno, R. *Building Construction Handbook.* Elsevier, Butterworth-Heinemann, Amsterdam, 2007.
2. Punmia, B. C., et al. *Building Construction: An Elementary As Well As Advanced for Engineering Students.* Laxmi Publications, 2016.
3. Iano, J., Allen, E. *Fundamentals of Building Construction: Materials and Methods.* Wiley. 2019.
4. Puybaret, E. *Sweet Home 3 D,* eTeks, 2016 (Open Access Software).

5 Building Finishes

UNIT SPECIFICS

Through this unit we have discussed the following aspects:

- The different types of floor finishes and their suitability.
- The process of laying and constructing the floors.
- The different types of plastering and their suitability.
- The process of painting and methods of application of paint on the surface.

For helping the students in the visualization of the content and to enhance their practical knowledge, the figures are provided in 2 D and 3 D.

This unit contains short and long answer-type questions along with the multiple choice questions, a list of references, and suggested reading so that one can go through them for practice. Some QR codes are given which can be scanned for more information on various topics of interest. The QR code provides extra knowledge about important topics.

RATIONALE

This unit on building finishes gives an idea about the finishing work of the different components of the building. It explains the different types of floor finishes and their suitability. A brief description of roofing materials and their suitability is also provided. The different types of plastering and their significance is also discussed in the detail. The content provided in this unit is important for planners, architects, and engineers.

Building finishing is an important requirement of a building. The knowledge of building finishes is important for civil engineers and construction planners. The students can implement the knowledge gained from this unit in the planning, construction, and designing of the buildings. The knowledge gained from this unit will help the students in the selection of suitable building finishes.

PRE-REQUISITES

Basic knowledge of building components is required for studying this unit.

UNIT OUTCOMES

The list of outcomes of this unit is as follows:

U5-O1: Identify the suitable type of floor finish.

U5-O2: Deciding the suitable roofing material for different buildings.

U5-O3: Explain the process and different types of plastering.

U5-O4: Understanding the precautions to be taken during the painting and plastering work.

Unit-5 Outcome	EXPECTED MAPPING WITH COURSE OUTCOMES (1 - Weak Correlation; 2 - Medium Correlation; 3 - Strong Correlation)				
	CO-1	CO-2	CO-3	CO-4	CO-5
U5-O1	1	-	-	-	3
U5-O2	1	-	2	-	2
U5-O3	-	-	1	-	3
U5-O4	-	-	1	-	3

5.1 FLOORS AND ROOFS

A floor may be defined as the flat surface provided at the plinth level and at the different elevations which are used for storing, living, and working. The floors divide the building into different levels. The topmost floor of a building is known as a roof. The floor beneath the ground level is called the basement floor while the floor provided at the plinth level is called the ground floor. The floors that are provided at various elevations are termed the first floor, second floor, etc.

5.2 TYPES OF FLOOR FINISH AND ITS SUITABILITY

A floor consists of two components i.e. base and flooring or floor finish. The base provides strength to the floor. It keeps the surface of the floor flat and prevents the settlement of the floor. The flooring is on the top portion of the floor and placed on the base. It provides smoothness to the floor. It is easy to clean and can be constructed using different materials like stone, timber, ceramic, etc. The different floor finishes used as flooring are explained in the upcoming sections.

5.2.1 Kota Floor Finish

Kota stones are a type of limestone that originated from the Kota region of Rajasthan. As Kota stones are locally available and possess sufficient strength, it is used as flooring material widely. These are fine-grained sedimentary stones consisting the silica and calcium carbonate. These types of flooring are shiny as well as durable. Kota flooring offers the following advantages:

(1) Kota flooring is smooth and shiny in nature. So, these are aesthetically pleasing.

(2) This type of flooring provides flexibility during finishing. It can be finished as per the requirement of the user.

(3) It resists moisture due to its low porosity. So, the Kota flooring can be used in regions having humid climates.

(4) Kota stone flooring is economical among natural stones.

Fig. 5.1: Kota finish

(5) These types of stones are available in a variety of colors.

Kota stone tiles are available in small sizes. So, there are numerous joints on the floor. These types of flooring require frequent polishing as these can get flaked with time. Figure 5.1 shows the Kota floor finish.

5.2.2 Marble Floor Finish

Marble stone is a type of metamorphic stone. When the composition of limestone is changed due to excessive pressure and temperature under the earth's surface, it is referred as a marble stone. Marble stones are widely used as a flooring material. Marble stones are white in color however blue, pink, yellow, and black shades may also exist due to the impurities present in the marble. These stones consist of calcium carbonate and which is alkaline in nature. Marble flooring has the following features:

(1) Marble is a naturally available stone with a very high ability to get finished. It gives a smooth and elegant appearance.

(2) Due to its smooth surface, the marble flooring is easy to clean.

(3) Marble flooring offers a range of colors. However, the color of the purest form of the marble stone is white. As it is a naturally occurring stone, the shades for each tile may vary slightly.

(4) These types of flooring remain cool but also offer good heat conduction. So, these can be used in cold regions too. Radiant heat can be used for warming the tiles.

(5) Marble flooring is highly durable. The floor remains stain free if maintained properly.

(6) Marble flooring provides continuity and the number of joints is less in the case of marble flooring.

Marble stone is a good choice for flooring material but it may remain slippery under moisture. Due to its slippery surface, it can prove hazardous to the residents and poses a risk of accidents. Scan the QR code to know more about different types of available marble flooring.

5.2.3 Granite Floor Finish

Granite stone is formed by the cooling of molten lava. It is composed of minerals like feldspar and quartz. It is nonporous in nature and durable as a flooring material. Granite flooring has the following features:

(1) The rigidity of the granite flooring is higher as compared to the other flooring materials.

(2) Granite flooring is durable in nature.

(3) The granite floor has sufficient resistance to heat. So, it remains cool during summer.

(4) Granite flooring can be used in damp regions as it offers resistance to moisture.

(5) It is easy to clean and requires low maintenance.

Fig. 5.2: Granite floor finish

(6) Granite flooring has a smooth surface.

Granite flooring is slippery due to the smoothness of the surface. Due to its heavy weight, it is difficult to handle. Granite flooring is difficult to install as compared to other flooring materials. Figure 5.2 shows the granite floor finish.

5.2.4 Ceramic Tiles and Vitrified Tiles

Ceramic tiles are composed of clay. The clay mixed with water is baked at a high temperature to manufacture the ceramic tiles. The ceramic tiles provide an earthen appearance. The tiles that are prepared by mixing 40% clay and 60% silica through the process of verification are known as vitrified tiles. The vitrified clay consists of clay, feldspar, quartz, and silica. Figure 5.3 (a) and (b) show the ceramic tiles and vitrified tiles respectively.

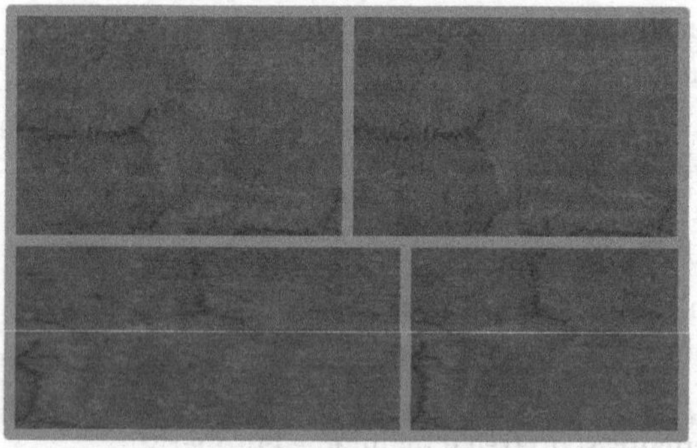

Fig. 5.3 (a): Ceramic tiles

Fig. 5.3 (b): Vitrified tiles

The vitrified tiles give a glassy appearance. The ceramic tiles and vitrified tiles have the following features:

(1) Ceramic tiles give a natural earthen appearance while vitrified tiles give an artificial glassy appearance.

(2) The texture of the ceramic tiles is rough as compared to the vitrified tiles. The vitrified tiles provide a smooth surface.

(3) The ceramic tiles are protected by the process of glazing while the vitrified tiles can be polished for providing a shiny surface.

(4) The flexural strength of ceramic tiles is lesser than vitrified tile. The flexural strength of the ceramic tiles is approximately 20 N/ mm^2 while the vitrified tiles have a flexural strength of around 35 N/ mm^2.

(5) Ceramic tiles are more porous in nature. Due to the high porosity of ceramic tiles, the frost resistance is lesser in these tiles as compared to the vitrified tiles.

(6) Ceramic tiles are more economical as compared to vitrified tiles.

(7) There is a requirement for skilled labor for the installation of ceramic tiles. These are more difficult to install as compared to the vitrified tiles.

(8) The breaking strength of the ceramic tiles varies from 700 N to 1000 N while the vitrified tiles have a breaking strength of more than 1100 N.

5.2.5 Chequered Tiles

Chequered tiles have grooves on their surface for making them non-slippery. The area of grooves should be at least 2% of the total surface area of the tile. Figure 5.4 shows the chequered tiles. The chequered tiles have the following features:

(1) The chequered tiles are durable. These are long-lasting and can withstand the action of wind and water.
(2) These tiles are easy to install. The grooves help in the interlocking of the tiles.
(3) The chequered tiles are non-slippery in nature. So, these can be used in the washrooms and on the carpet.
(4) These types of tiles possess high strength.
(5) The chequered tiles are economical.

Fig. 5.4: Chequered tiles

5.2.6 Paver Blocks

The paver blocks are prepared by mixing the cement and sand in different rations. These blocks are prefabricated in the factories. The strength of the paver blocks depends on the mix ratio of cement and sand. These tiles are manufactured in different shapes like rectangles, hexagons, L-shape, and X-shape. Figure 5.5 shows the paver blocks. The paver blocks have the following salient features:

(1) The paver blocks are very durable and can last up to 20 years.
(2) These can be interlocked with each other. Hence, these are easy to install.
(3) The paver blocks are very strong as these can take very heavy wheel loads.
(4) The paver blocks require less maintenance and cleaning.
(5) The paver blocks are easy to remove and reinstall. If one block is damaged, it can be replaced.

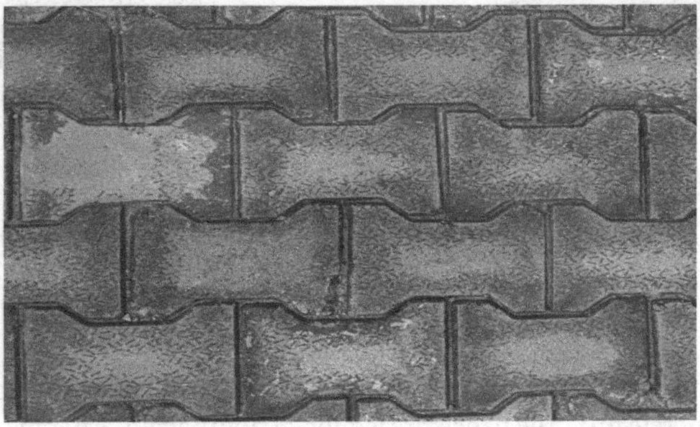

Fig. 5.5: Paver Blocks

5.2.7 Concrete Flooring

Concrete flooring is widely used in residential and public buildings. The concrete floors are strong, durable, and relatively economical. The concrete floor is constructed in three layers as shown in figure 5.6.

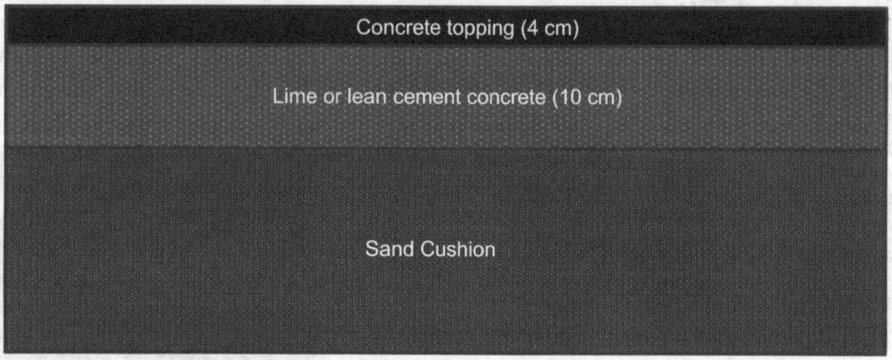

Fig. 5.6: Concrete flooring

After the preparation of the base, the area is divided into rectangles or squares with the help of wooden panels. The size of the sides of the squares or rectangles should not be more than 1.5 m. The height of the battens should be equal to the depth of the floor. M 15 or M 20 concrete mix is used for the construction of the concrete floor. The alternative panes are filled with concrete mix. The surface is leveled with the trowel and straight edge. The panels filled with concrete are left for 24 hours. Figure 5.7 shows the alternate filling of the panels.

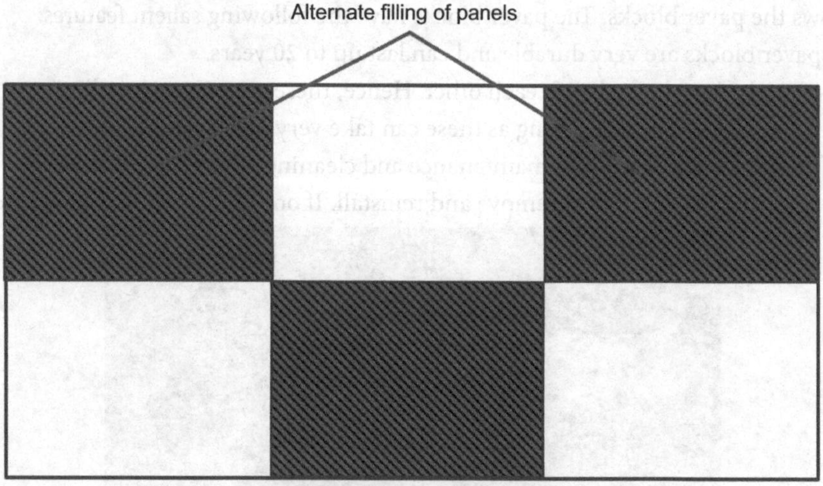

Fig. 5.7: Alternate filling of panels

As the concrete in the alternative panels is set, the formwork of battens is removed and the unfilled panels are filled with concrete mix. The panels are left for 24 hours again. After the setting of the concrete, the battens are removed. The curing of the concrete floor should be done with water in a proper way. The concrete flooring has the following salient features:

(1) The concrete floors have sufficient strength. These are durable in nature.
(2) The concrete flooring offers good resistance to dampness.
(3) These floors offer good resistance to fire.
(4) The concrete floors can be used for any loading conditions. These can sustain very heavy loads.

(5) These floors require less maintenance and can prove cost-effective.

Concrete floors are economical in the long run however the initial cost of construction of concrete floors is high. If the construction is defective, the repair costs for the concrete floors will be high. So, these types of floors require skilled workmanship.

5.2.8 Wooden Flooring

Timber flooring is used widely in hotels, gymnasiums, movie halls, etc. In India, timber flooring is used normally in hilly regions due to the humid climate. The wooden floors don't remain in direct contact with the ground. These are kept 1 m to 1.5 m above the ground to protect them from dampness. The wooden flooring has the following advantages:

(1) Wood is a light material that is easy to transport and install.
(2) The wooden floor installation is quick due to its easy handling and placement.
(3) The wooden floors remain free from dampness.
(4) These types of floors can absorb vibrations and shocks. So, these are recommended in earthquake-prone regions.
(5) The timber flooring is aesthetically pleasing and stylish.

The wooden flooring has the following limitations:

(1) The timber flooring is less durable.
(2) It requires frequent maintenance.
(3) Wood is prone to termites and other insect attacks.
(4) Wooden floors are costly in manufacturing.
(5) This type of floor doesn't offer good resistance to fire.

5.2.9 Skirting and Dado

Skirting tiles are provided along the perimeter of the bottom of the wall. These tiles connect the floor and wall. Skirting is a kind of wall treatment and it is provided just above the floor with a height of 7.5 cm to 10 cm. In usual cases, these are the same as the floor tiles. Figure 5.8 shows the skirting tiles. The skirting tiles are provided for the following purposes:

(1) Skirting tiles may be used to cover the exposed electric cables.
(2) The unintentional gaps between the wall and floor are covered with the help of skirting tiles.
(3) The skirting tiles provide a gap between the wall and the furniture. So, these tiles protect the wall from scratches due to the furniture.
(4) The skirting tiles are aesthetically pleasing and enhance the beauty of the floor and walls.

Fig. 5.8: Skirting tiles

The dado tiles are provided in the kitchen and washrooms. These are provided above the cabinet and slabs in the kitchen. The dado tiles protect the wall from the stains of food and liquid. These are easy to clean. The dado tiles also protect the wall from the water.

5.3 PROCESS OF LAYING AND CONSTRUCTION

The floor has two components i.e. base and flooring. The base provides structural strength to the floor while the flooring provides a smooth surface. The process of laying and construction of the floor is explained in the following steps:

(1) The construction of the wall must be completed before the beginning of the construction of the floor.

(2) The area of the floor is cleaned. The roots, vegetation, and any other irrelevant material is removed properly. If the depressions are created by the removal of vegetation, these must be filled and compacted with the soil. Clean sand and soil are filled in the floor area.

(3) The reference marks above 15 cm of the floor level are marked on the walls. These marks are used as a reference while constructing the floor.

(4) The soil spread on the floor according to the required slope. The soil is properly compacted to avoid settlements.

(5) If the soil used in the construction of the floor is weak, it can be treated. The strength of the soil can be enhanced by the chemical or physical treatment of the soil.

(6) The layer of lime concrete or cement concrete of a depth of 10 cm to 15 cm is placed on the floor. The concrete is compacted and rammed properly.

(7) The curing by sprinkling the water on the concrete is done after setting the concrete.

(8) After setting, the base layer is cleaned and flooring is provided. The base must be cleaned properly before providing the topping.

5.4 FINISHING AND POLISHING OF FLOORS

The surface of the floor loses its shine with time. So, the floors need to be maintained periodically. The floors are polished and finished manually or mechanically. The finishing and polishing of the floors offer the following benefits:

(1) The polishing of the floors removes the marks of dirt and stain from the floor. The polishing and finishing make the floor smooth.

(2) The polishing removes the dullness of the floor.

(3) The finishing and polishing restore the shine and smoothness of the floor which can be reduced with time.

Polishing pads installed in the floor polishers are used for polishing the floors. The polisher equipment is used for scrubbing, stripping, and polishing of hard surfaces like marble or concrete floors. The pads can be used manually for the removal of moderate defects. The grinding machine with heavy-duty motor and disc revolving at the bottom. The grinding pads or polishing pads are made of carborundum stones. These are manufactured in sizes of 60 grits to 400 grits. To see the video of the polishing of the floor, scan the QR code.

SCAN ME

5.5 ROOFING MATERIALS

The permanent covering provided on the rooms to protect them from the impact of weather, sun, rain, storms, etc. is called a roof. It is the topmost part of a building. The roof is constructed with trusses and roof covering. The tiles, sheets and slabs are usually used for the construction of the roof in a building. The different materials that are used for covering the roofs are enlisted as follows:

- Reinforced cement concrete (RCC)
- Mangalore tiles
- AC Sheets
- Corrugated GI Sheets
- Plastic sheets
- fiber sheets

5.5.1 Reinforced Cement Concrete (RCC)

Reinforced cement concrete (RCC) slabs are widely used as roofs in residential and public buildings. RCC can be designed according to the superimposed loads. If the span of the roof is more than 3 m, beams can also be provided to support the slab. RCC slabs offer sufficient resistance to fire and dampness. The RCC slabs are durable and long-lasting. The RCC slabs are laid in the following three steps:

(1) At the initial stage, the concrete requires support after poring. So, the formwork is erected in the first step. The formwork should be provided according to the provisions of Indian Standard code IS: 14687-1999.

(2) The concrete is strong in compression but weak in tension. So, the reinforcement is designed for the concrete. The steel bars are provided at the designed spacing in the second step. The deformed steel bars are provided according to the provisions of the Indian Standard code IS: 1786-1985.

(3) The concrete mix is designed according to the codal provisions. M 15 or M 20 concrete mix can be used in the construction of roofs. The concrete mix is poured into the formwork and compacted with the help of vibrators. After setting the concrete, it is cured by sprinkling the water on the surface for 14 days to 21 days. A bitumen layer is provided on the concrete layer and a lime concrete layer is provided at the top. The bitumen layer acts as a damp proof course (DPC) in the roof and prevents the entry of moisture. Figure 5.9 shows the RCC slab as a roof.

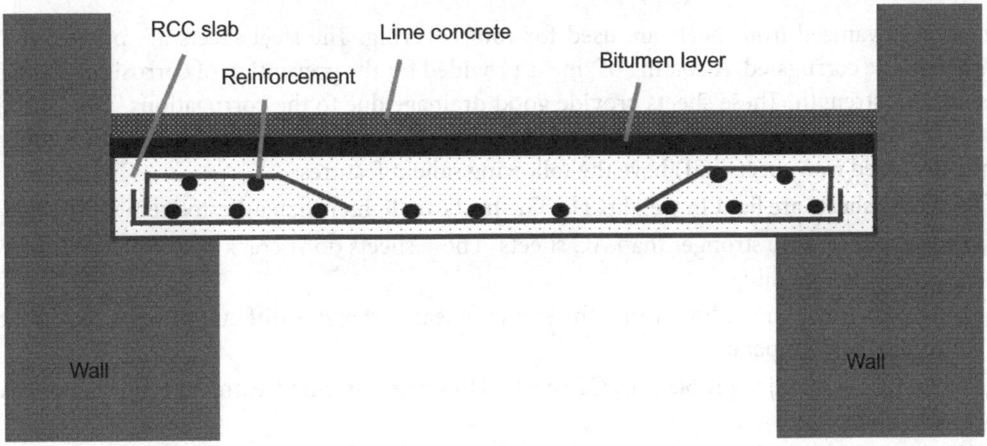

Fig. 5.9: RCC slab

5.5.2 Mangalore Tiles

Tiles as a roofing material are used widely. A tile is made in two parts. The bottom part is flattened while the top part is semi-circular in shape. The local tiles are improved by modern manufacturing models. The Mangalore tiles are manufactured in the 1860s first time. Georg Plebst identified the clay that is used for making the Mangalore tiles. As the raw material for Mangalore tiles is discovered in Mangalore city of India, these are named on the basis of their origin city. The fine clay is filled in the mould and baked at a high temperature. The tiles are glazed for a shiny appearance. The Mangalore tiles have the following salient features:

(1) The Mangalore tiles are cost-effective.
(2) These tiles are light in weight and can be lifted easily.
(3) These tiles offer good resistance to fire.
(4) The Mangalore tiles can be moulded into any desired shape.
(5) These tiles look aesthetically pleasing.
(6) These tiles have less strength. These tiles are unable to withstand heavy loads.

5.5.3 Asbestos Cement (AC) Sheet

The asbestos cement sheets are prepared by mixing 15% of the asbestos fiber into the cement. These sheets are used as the sheds of factories and workshops. The AC sheets have the following salient features:

(1) These sheets remain less hot in the summer season. So, it is easy to work under them.
(2) The AC sheets offer good resistance to fire.
(3) These sheets prevent the entry of water. These are completely waterproof.
(4) The AC sheets don't rust with time.
(5) These sheets require no paint.
(6) These sheets can get cracked. So, the AC sheets should be handled carefully during the laying process.

5.5.4 Galvanized Iron (GI) Sheet

Corrugated galvanized iron sheets are used for roof covering. The steel sheets are pressed by groove rollers and made corrugated. A coating of zinc is provided for the prevention of corrosion. The GI sheets have sufficient strength. These sheets provide good drainage due to the corrugations. The GI sheets are generally 0.9 m wide. These are manufactured in the length of 1.8 m, 2.2 m, 2.5 m, 2.8 m, 3.0 m, and 3.2 m. The galvanized iron sheets (GI) have the following salient features:

(1) These sheets are light in weight. The GI sheets are lighter than asbestos cement sheets.
(2) These sheets are stronger than AC sheets. These sheets don't crack due to the collision during lifting and handling.
(3) These sheets remain hot during the summer season. So, it is difficult to work under them due to the high temperature.
(4) Rusting is a major problem in GI sheets. However, the zinc treatment reduces the rusting of these sheets.
(5) These sheets don't provide sufficient resistance to sound and heat.

5.5.5 Plastic and Fiber Sheet

Plastic is a versatile material that is used as a roof covering. It protects the building from ultraviolet (UV) rays. The plastic sheets are light in weight. The plastic sheets are used as sheds for parking, farmhouses, garages, etc. The plastic roofing sheets have the following salient features:

(1) The plastic sheets are durable. These can resist the impact of weather, rain, heat, etc.
(2) These sheets are light in weight. So, the transportation and placement of plastic sheets are comparatively easy.
(3) These sheets look aesthetically pleasing. These can be printed with various designs.
(4) The plastic sheets can be customized according to the requirement of the user.
(5) The plastic sheets can provide entry to the light through them.
(6) The plastic acts as an insulator of the heat.
(7) These sheets are cost-effective and prove economical.

The characteristics of the plastic can be improved by providing fiber reinforcement. The sheets that are reinforced by vinyl, epoxy, etc. are known as fiber-reinforced polymer (FRP). The fiberglass sheets give more natural light. The strength of the fiber sheets is more than the normal plastic sheets. In regions of cool temperatures, fiber sheets can be used. To know more about the roofing materials, scan the barcode.

SCAN ME

5.6 TERMS USED IN ROOFS

The various terms used in the roofs are given as follows:

(1) **Truss:** In the pitched roof, a frame of a triangular shape is used to form the base of the roof. This frame is known as a truss. A number of trusses are used for the construction of a pitched roof.
(2) **Span:** The horizontal distance between the two walls is known as the span of a truss.
(3) **Pitch:** The angle of the roof with respect to the floor is known as the pitch of the roof. If the pitch of the roof is equal to or less than 10°, it is called a flat roof. If the pitch is more than 10°, it is called a sloped or pitched roof. The slope of the pitched roof is kept from 25° to 60°.
(4) **Ridge:** The upper part of the roof where the slopes of both sides of the roof meet is known as a ridge.
(5) **Rise:** The rise of the roof is the vertical distance between the top of the ridge and the horizontal line connecting the lower ends of the truss.
(6) **Principal rafter:** The top inclined member of the truss on which purlins are placed is known as the principal rafter. The principal rafters can be seen in Figures 5.11 and 5.12.
(7) **Tie beam:** The bottom horizontal beam that holds the side rafters is known as a tie beam.
(8) **Strut:** The struts are the members that are placed between the tie beam and principal rafters for binding them.
(9) **King post:** The vertical structural member in the middle of the king post truss that lies between the tie beam and the ridge is known as a king post.

(10) **Queen post:** These are two or more vertical members that hold the tie beam and the principal rafter. These are provided at the central part of the queen post truss.

(11) **Eaves:** The ends of the truss that projects beyond the wall are known as eaves.

(12) **Purlins:** The purlins are used for fastening the two rafters.

(13) **Cleat:** The purlins are provided at a slope. Cleats are provided for preventing the purlins from displacement.

(14) **Battens:** Flat and long strips that are made of wood are known as battens.

5.7 TYPES OF ROOFS

The type of roof is selected based on the loading, weather, the distance between walls, etc. The roofs can be classified into two types:

(1) Flat roof

(2) Pitched roof

5.7.1 Flat Roof

A flat roof is a type of roof that has a flat upper surface. The angle of inclination of a roof is known as pitch. The flat roofs have a maximum pitch of 10°. In high-temperature regions, the roof is used for sleeping and other work. So, flat roofs are frequently used in India. The flat roofs are exposed to sunlight and rain directly. Due to variations in temperature, these roofs get contracted and expanded. So, these are cracked due to the expansion and contraction phenomenon. Hence, damp proofing and heat-resistant treatment are necessary for such roofs.

Battened flat roofs are cost-effective flat roofs. The roof is constructed with a thick layer of mud and clay tiles placed on it. It is also known as a mud roof. It is used in low-rain areas. The timber joists are used for supporting the battened roof. So, the age of the roof depends on the age of the timber joists. As the timber can be affected by moisture and dampness, the joists can be coated with bitumen.

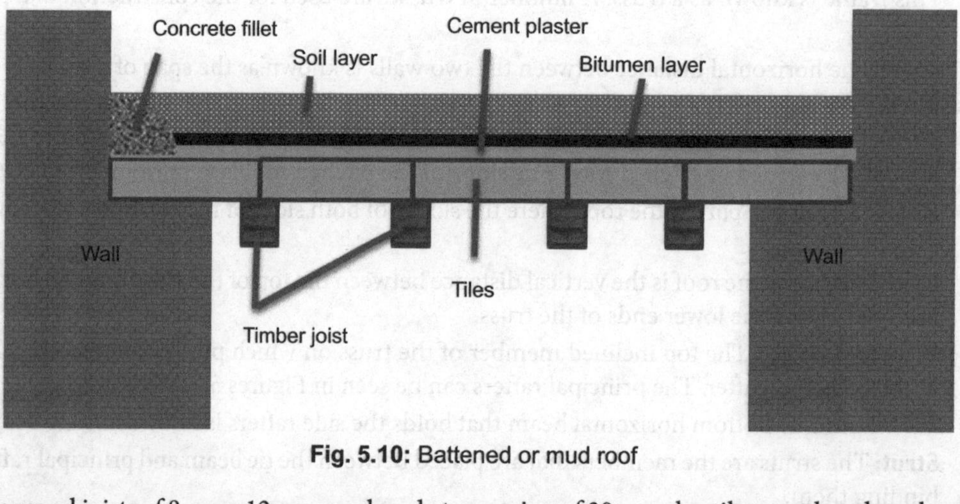

Fig. 5.10: Battened or mud roof

The wood joists of 8 cm × 12 cm are placed at a spacing of 30 cm. clay tiles are placed on these joists. The size of the clay tiles is generally 30 cm× 15 cm× 5 cm. Cement mortar of 1:3 mix is used for placing the tiles. A layer of cement plaster is placed above the tiles. A bitumen layer as a damp proofing treatment is provided above the cement plaster layer. At the top, a 10 cm layer of soil is placed. The slope of the roof is kept at 1 in 40. This slope is provided for the drainage of rainwater. RCC battens may be used instead of timber joists. These offer better resistance to fire. The RCC battens are also more durable and long lasting as compared to the timber joists. Figure 5.10 shows the battened or mud roof.

The flag-stone roofs are used in the regions where flag-stone is available locally. These roofs prove economical in the regions of Haryana and Rajasthan. The flagstone roofs are cost-effective and easy to construct. The flagstones of 2 m to 2.5 m in length are placed to form the roof. However, if the span of the room is large, T-section steel girders can be used for supporting the stone sections. The joints of the stone are sealed with lime mortar. A 10 cm thick lime concrete layer is placed on the top. The layer is finished and the roof is sloped for drainage purposes. RCC slabs are also used as roofs. The details of the RCC slab as a flat roof are described in the previous section.

5.7.2 Pitched Roof

If the slope of the surface of the roof is more than 10°, it is known as a pitched roof. A pitched roof is constructed with the help of trusses. These are also called sloped roofs. The pitched roof is light weighted and damp-proof. For long spans, pitched roofs are used. These roofs can't be used for sleeping or standing. The components of the trusses are changed based on the span.

5.7.2.1 King Post Truss

King post truss is created by the combination of triangular frames. A king post truss consists of the members like tie beams, rafters, struts, king posts, and ridge beams. The rafter and ties need support as these can get deformed under the impact of their self-weight.

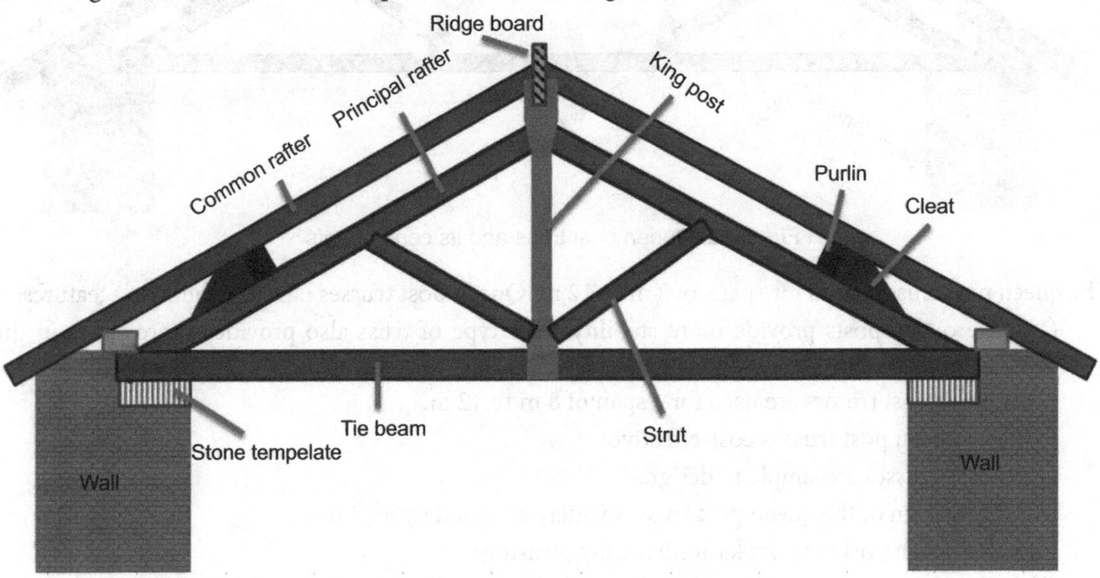

Fig. 5.11: King post truss and its components

So, additional rafters are provided to support the upper members in the middle of the truss. The king post truss can be used for shorter spans. The maximum span for the king post truss can't be more than 9 m. A vertical column is placed in the mid of the truss. This column is known as the king post. Figure 5.11 shows the king post truss and its components. The king post truss has the following features:

(1) The king post is used for short spans. It can be used for the construction of bridges too.
(2) These are used in the construction of timber roofs.
(3) The king post truss is cost-effective.
(4) Due to the number of structural members in the truss, the intermediate space can't be used.
(5) King post truss is not suitable for spans more than 9 m.

5.7.2.2 Queen Post Truss

If the span is larger than 9 m, the king post can't be used. In this case, a queen post truss is used. The king post trusses can bend due to their own weight. Two queen posts are provided to prevent the bending of the truss. The horizontal beam at the top of the queen post is called a straining beam. The gap is separated by the purlin and cleat. Figure 5.12 shows the queen post truss and its components.

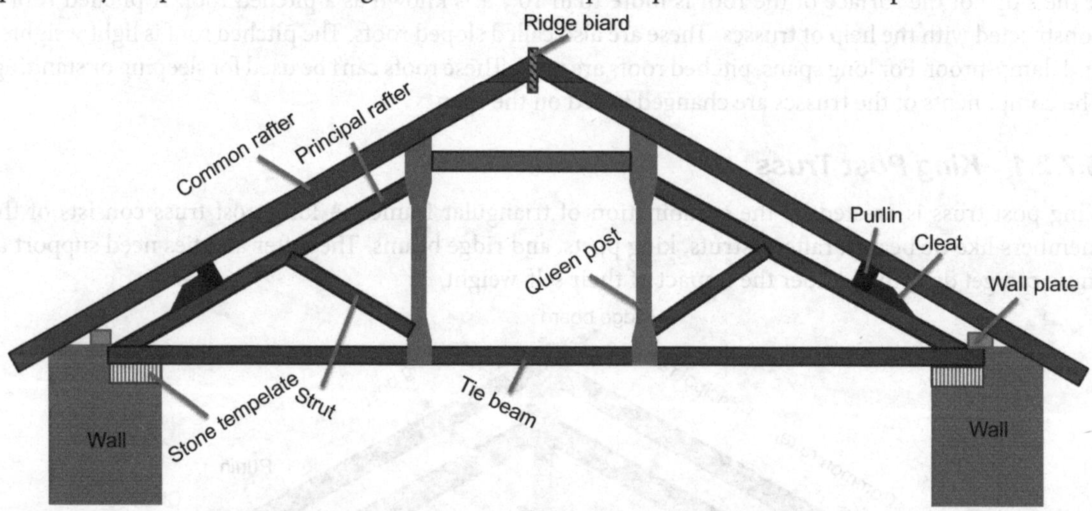

Fig. 5.12: Queen post truss and its components

The queen post truss is used for spans of 8 m to 12 m. Queen post trusses have the following features:

(1) The queen posts provide more stability. This type of truss also provides more space in the room.
(2) Queen post trusses are used for a span of 8 m to 12 m.
(3) The queen post truss is cost-effective.
(4) These trusses are simple to design.
(5) The design of the queen post truss is widely accepted by engineers.
(6) The queen post truss looks aesthetically pleasing.
(7) This type of truss is used in churches and other buildings that require more open space.
(8) The queen post truss is not used for spans of more than 12 m.
(9) This type of truss is not suitable for regions with high rainfall.
(10) The queen post truss performs poorly under the uniformly distributed load.

5.8 PLASTERING AND ITS NECESSITY

Plastering is a procedure in which a wet mixed material (cement/lime mortar) is used to cover the internal and external walls and rough patches on the surface of columns, beams, and other building parts. It provides even, clean and durable surfaces for easy and direct decoration. Plastering is done on buildings components for the following reasons:

- It protects the external surface of the building from environmental effects such as rain and wind.
- It provides a smooth and even surface in which dust or mud cannot stay on the wall.
- It provides a better surface for decoration.
- It helps in protecting building surfaces from bugs or insects.
- It covers the defects due to unskilled workmanship.
- It provides an excellent base for whitewashing, painting, or distempering.

5.9 PROCEDURE OF PLASTERING

The plastering is performed in various steps. The process of plastering is given as follows:

(1) **Surface preparation:** Before beginning the plastering, the surface is prepared for taking the coats of plaster. The surface is prepared as follows:
   - ➢ All mortar joints have to be cleaned up to a depth of 10 mm in the brick mortar and 15 mm in the stone mortar for better plastering.
   - ➢ Wire brushes are used to remove mortar dusting over the wall surface.
   - ➢ Any path holes are filled and unevenness in the surface is leveled.
   - ➢ The clean surface should be washed and kept wet before use for better suction.
   - ➢ If any projection present on the wall surface is more than 12 mm, then it has to be removed for obtaining a uniform surface.
   - ➢ If plastering is to be done on an old wall surface, all dust, paints, oil, grease, etc. should be removed.

(2) **Groundwork before plastering:** The groundwork before plastering is done as follows:
   - ➢ To get plastering of uniform thickness, the wall surface is marked with dots. Dots refer to plastering a small patch of size 15 mm x 15 mm with a thickness of 10 mm.
   - ➢ These dots are fixed horizontally and vertically at a distance of 2 m between them from center to center.
   - ➢ Plumb bobs are used to check the vertical of dots.

(3) **Base coat application:** The base coat application is done in the following steps:
   - ➢ The thickness of base coat plastering in brick masonry is kept around 12 mm and in concrete masonry, it is 9 mm.
   - ➢ In bathe se coat, the ratio of cement to sand is 1:3 to 1:6
   - ➢ Flat wooden floats are used to level the surface

(4) **Finish coating application:** The finish coating is applied in the following steps:
   - ➢ The finishing coat thickness varies from 2 to 3 mm.
   - ➢ The ratio of cement to sand is 1:4 to 1:6.

- ➢ Flat wooden floats and steel trowels are used for even surfaces and provide the finishing touch.
- ➢ Application is done from top to bottom and in one single operation so that no cracks can be developed.

(5) **Curing after plastering:** After finishing the coat, the water sprinkling is done for at least 7 days to gain proper strength and hardness. Plastering works is raped with jute gunny bags to keep wet for a longer duration. Proper curing work should be carried out otherwise cracking formation can occur.

5.10 DOUBLE COAT PLASTERING

The double-coat plastering is done in a particular sequence. The steps followed for double coat plastering are explained as follows:

- Raking of mortar joints is done up to a depth of 20 mm
- Cleaning the surface with a wired brush and then watering the surface
- The uneven surface is leveled by filling the potholes
- The first coat is applied, which is also known as the rendering coat. To maintain the uniform thickness of the plastering dots are made on the surface of size 15 cm x 15 cm at a distance of 2 m apart in the vertical and horizontal directions. Vertical dots are joined by making vertical lines of mortar and thickness is checked by a plumb bob. This vertical line of mortar is known as a screed. Mortar is then filled inside this screed and the surface is properly leveled and finished
- Mechanical keys are formed over the rendering coat before it attains hardness.
- The rendering coat is watered for 2 days and left to attain complete dry
- The second coat is also known as the finishing coat and may have a thickness varying from 2 -3 mm. The rendering coat is damped before applying the finishing coat. A finishing coat is applied from top to bottom in one action to dispose of joint marks.

5.11 SINGLE COAT PLASTERING

Single-coat plastering is done where the work is not important or of inferior quality. This coat is applied the same way as the double coat plaster except for that rendering coat. Only the finished coat is applied in the case of single-coat plastering.

5.12 ROUGH FINISH AND NEERU FINISH

In the rough finish, coarse aggregate is mixed with cement and sand to form finishing mortar. The ratio in which cement, sand, and coarse aggregate are mixed is 1:1.5:3. The size of coarse aggregate varies from 3 m to 12 mm. Large towels are used to dash the mortar against the prepared plastered surface. Wooden floats are used for rough finishing. This type of finish is water-proof, resistant to cracking of joints, and used as renderings for the external surface.

In Neeru finishing, lime paste or putty is used to render the cracks and level the uneven surface to obtain a smooth and leveled finish. The minimum thickness of the Neeru coat is 3 mm. Undercoat or base coat plaster should be in the ratio of 1:4 for Neeru finishing. Lime used for Neeru finishing should be slaked at the site before use. Neeru coat is applied instantly after the base coat has attained the initial setting time i.e. about 4 hours.

5.13 PLASTER OF PARIS (POP)

Plaster of Paris is made by heating finely powered gypsum at 160°C - 170°C. It is white in color. It starts to get settle within 3 - 4 minutes of adding water. So, retarders are added in POP to delay the setting time. POP is used in combination with lime for repairing potholes and cracking on the surface finish and for ornamentals work. POP is also known as gypsum plaster. It has the following properties:

- POP can be used adequately over timber material and components made up of metals as it has fire-resistance properties.
- It is very light in weight.
- It acts as a sound insulator.
- Ornamentals works carried out in building use effectively POP for construction
- It can be easily attached to fibrous materials.
- Its setting time is less and there is no change in volume after set, so there is no shrinkage on drying.
- It can only be used for interior work in building as gypsum plaster is soluble in water.

5.14 SPECIAL PLASTERS

The plaster that is prepared for specific purposes is known as a special plaster. Some of the special plasters are discussed as follows:

- **Acoustic plaster:** In this type of plastering gypsum is added to the final coating in finishing. The mixing of gypsum results in a chemical reaction that produces gaseous bubbles resulting in the formation of a small opening in the final coating. These tiny openings help in absorbing sound. Such plastering is done on the interior walls of the hall, auditorium, etc. Two coats are applied with each thickness of 6 mm.
- **Barium plaster:** This type of plaster is made up by mixing barium sulfate with cement and sand. It is used in X-rays room to protect persons working there from harmful X –rays.
- **Keene's cement plaster:** This plaster can be made by calcinating POP with alum. It provides a glass-like polishing surface. It is extensively used in the construction of ornamentals and decorative works.
- **Asbestos marble plaster:** These plasters are obtained by mixing asbestos with cement and finely crushed marble. It imparts marble-liking finishing.

5.15 STUCCO PLASTER

Stucco plaster is made of aggregates, water, and binder. The binder can be lime or cement. In conventional times, the stucco plaster is created from lime while in modern times it is prepared with cement. After hardening, stucco plaster forms a very dense layer. The stucco plaster may also be used for decorative purposes. The stucco plaster has the following features:

(1) The stucco plaster looks aesthetically pleasing. So, it can be used for decorative coating.
(2) It can be used on both sides of the wall.
(3) The thickness of the stucco plaster is 25 mm generally. It is applied in three layers.

5.16 SPONGE FINISH

The sponge finish is also called sand finish. This is done in two coats. In the first layer, the cement sand is taken in the ratio of 1:4. The first layer is provided with a thickness of 12 mm. The first layer is cured by sprinkling the water for 7 days. After hardening the first layer, a layer with a cement-sand ratio of 1:1 is provided. It is generally 8 mm thick. The layer is finished with the sponge to provide the required surface characteristics. The sponge helps in getting a smooth render finish as shown in figure 5.13.

Fig. 5.13: Sponge finish

5.17 PEBBLE FINISH

In the pebble finishing, the small pieces of stones having diameters from 10 mm to 20 mm are dashed on the surface of the wall. The mortar used in the plastering has a cement-sand ratio of 1:3. The pebbles should be washed with water properly to remove the impurities. The pebbles are pressed with the trowel or straight edge in the wall. Figure 5.14 shows the pebble finish. The pebble finish offers the following features:

(1) The pebble finish offers a rough surface.
(2) The pebble finish is aesthetically appealing.
(3) A pebble finish is used for the exterior walls.
(4) This type of finish is durable in nature.
(5) The pebble finish requires less maintenance. The finish must be washed with water regularly. After five years, the finish is sealed to maintain its beauty.
(6) The pebble finish is cost-effective and proves economical.
(7) The material is prone to degradation with time.

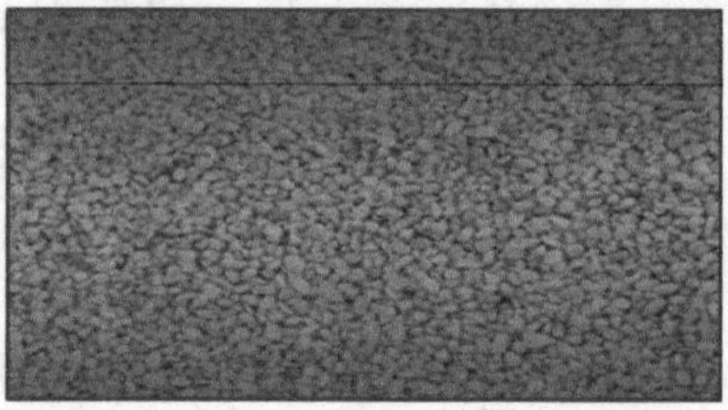

5.18 PLASTERBOARD

The plasterboard is made of gypsum. These are used for dry construction as an alternative to cement concrete construction. Plasterboards are invented in 1940. These are used for the construction of walls and ceilings. The sheets made of gypsum are fastened to the frame of the house with the help of screws, nails, or adhesives. Figure 5.15 shows the plasterboards.

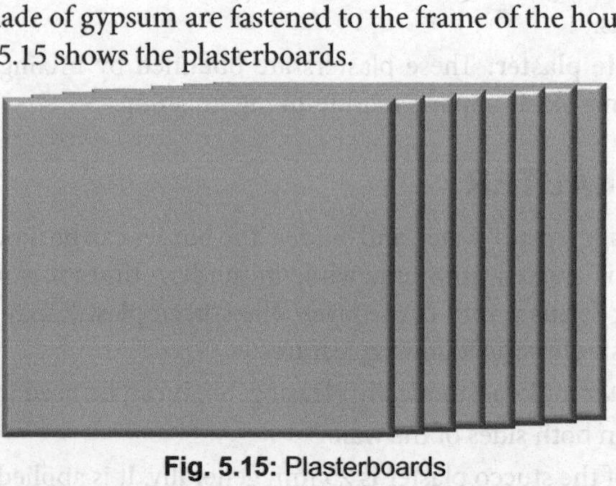

Fig. 5.15: Plasterboards

The plasterboard consists of a surface paper that makes the surface of the wall smooth. The plasterboards have the following features:

(1) The plasterboards are easy to install. This type of construction doesn't require mortar for adhesion. The plasterboards can be fixed with the help of fasteners.
(2) Plasterboards are cost-effective materials. The cost of the plasterboards is less than the cement and brick.
(3) This type of construction offers good resistance to sound. The plasterboards act as a barrier to sound.
(4) The material used in the making of plasterboards is environmentally friendly. It doesn't contain harmful chemicals.
(5) These boards are fire-resistant and moisture-resistant.

5.19 WALL CLADDINGS

Wall cladding is the process of providing the layer of one material over the layer of other material to protect the building components from the impact of moisture. The cladding can be provided on the wall by providing a layer of any material. The wall claddings can be made of stone, vinyl, fiber cement, wood, steel, etc. It is an extra protective layer that act as a decorative surface. The wall cladding has the following advantages:

(1) Cladding provide an attractive surface to the interior and exterior walls.
(2) The wall cladding prevents the penetration of the moisture.
(3) The cladding is resistant to the stains and scratches.
(4) These can be washed easily.
(5) It protects the exterior wall from the impact of weather.
(6) The wall cladding acts as a good thermal insulator.
(7) It acts as a barrier to the sound. Hence, it reduces the noise coming from the outside of the building.
(8) The wall claddings offer good resistance to fire. Scan the QR code to know more about the wall cladding.

5.20 PRECAUTIONS TO BE TAKEN IN PLASTERING

The plastering work should be performed carefully. The different precautions that should be taken during the plastering work are enlisted as follows:

(1) The material for plastering should be prepared by considering the provisions of the Indian standard codes.
(2) The cement used in the plastering should preferably be low-heat cement. Quick hydration of the cement can induce cracking in the surface.
(3) The surface of the wall should be prepared before beginning the plastering work. The loose mortar should be scrapped off with the help of brushes.
(4) The surface should be soaked with water before the plastering work.
(5) In the case of the exterior wall, the layer of plastering should be 18 mm thick. It can be applied in two coats. The first coat of 12 mm is applied and the second coat of 6 mm is applied.

(6) Straight edge and plumb bob should be used for checking the verticality of the plastering.
(7) The surface of the wall should be cured by sprinkling the water daily for 10 to 12 days after plastering.

5.21 DEFECTS IN PLASTERING

If the plastering is not done properly or the provisions of the code are not followed, the plaster can get defective. The various defects of plaster are enlisted as follows:

(1) Blistering
(2) Cracking
(3) Plaster falling
(4) Efflorescence

(1) **Blistering:** The blistering defect occurs in the lime plaster. Sometimes, the lime in the lime plaster is not calcinated completely. So, the unslaked lime particles remain in the paster that absorbs the moisture from the environment. Due to the absorption of moisture, the unslaked lime particles get calcinated. The surface of the plaster bulks up and the plaster starts crumbling. The blistering can be prevented by calcinating the lime properly before using it in plastering.

(2) **Cracking:** The plaster can be cracked due to multiple reasons. If the surface is not prepared well before the plastering work, the plastering can get cracked. The cracking in plaster may also occur due to frequent changes in the temperature. The variation in the temperature results in the shrinkage and expansion of the plaster which leads to the cracking of the surface. Sometimes, the excessive thickness of the plaster layer may be the reason of the cracking of the plastering. The cracking can be prevented by improving the quality of plaster work. The thickness of the plaster layer is kept between 15 mm to 20 mm.

(3) **Plaster falling:** The plaster falls off the wall due to various reasons. The primary reason for the falling of the plaster is poor bonding between the layers of the plaster or wall. It may be due to the presence of moisture at the surface of the wall. If the binding material like cement is not provided in the proper ratio, it can lead to poor bonding that results in the falling of the plaster. The falling of the plaster can be prevented by preparing the surface properly before proving the plaster layer. The amount of binding material should be provided according to the mix design.

(4) **Efflorescence:** If the amount of salt in the bricks or water used in the construction is high, it may appear on the surface of the plaster. This phenomenon of appearing of salt on the surface of the wall or plaster is known as efflorescence. The efflorescence can be prevented by using good-quality brick. The damp proof course provided at the plinth may also help in the prevention of efflorescence.

5.22 POINTING

The process of coating the joints with mortar is known as pointing. In pointing, the whole wall is not plastered but only the joints are coated with mortar. The wall need not to be painted in this case. So, it provides a natural appearance.

5.22.1 Necessity of Pointing

The joints in the wall are the weakest part. The water penetrates through the joints. So, the joints are sealed with the help of mortar in the case of pointing. The pointing increases the strength of the joints.

It seals the joints and protect them from the impact of rainfall and sunlight. Pointing is used for making the inferior masonry strong. For example, the mud masonry is weak. So, the joints can be protected by pointing in this case.

To sustain the natural appearance of the wall, pointing can be used. In the case of plastering, the whole wall is covered while pointing maintains the natural beauty of the wall. The pointing work is more durable than the plastering work as pointing cuts off the moisture. These work are also cost-effective due to the minimal use of the mortar.

5.22.2 Types of Pointing

The pointing is provided in different types. The selection of pointing depends on the texture, location of the wall surface, aesthetics, and outlay. The different types of pointing are discussed as follows:

(1) **Flush pointing:** Flush pointing is the simplest form of pointing. In this type of pointing, the mortar is filled in the joint and flattened with the help of a trowel.

(2) **U-grooved pointing:** The jointers are used to make U-shaped grooves that are filled with mortar. This type of pointing looks good aesthetically.

(3) **V-grooved pointing:** The V-shaped grooves are made by jointers and filled with mortar. These pointings are used in the case of stone masonry.

(4) **Keyed pointing:** In keyed pointing, the semi-circular groove is filled with mortar. It is used in brick masonry generally.

(5) **Raked pointing:** In raked pointing, the square grooves are made for a stylish appearance.

(6) **Weathered pointing:** In this pointing, a slope is provided. This slope helps in draining the water rapidly from the joint.

(7) **Mason-**V pointing: Mason-V pointing is used in stone masonry. The mortar points out in this case.

(8) **Bastard pointing:** In the bastard pointing, a strip of 6 mm width is extended from the wall by 3 mm.

(9) **Beaded pointing:** In the beaded pointing, a semi-circular strip extends beyond the wall.

(10) **Tuck pointing:** In the tuck pointing a rectangular groove is made. This groove is filled with putty of different colors that make the pointing beautiful. Figure 5.16 shows the different types of pointings.

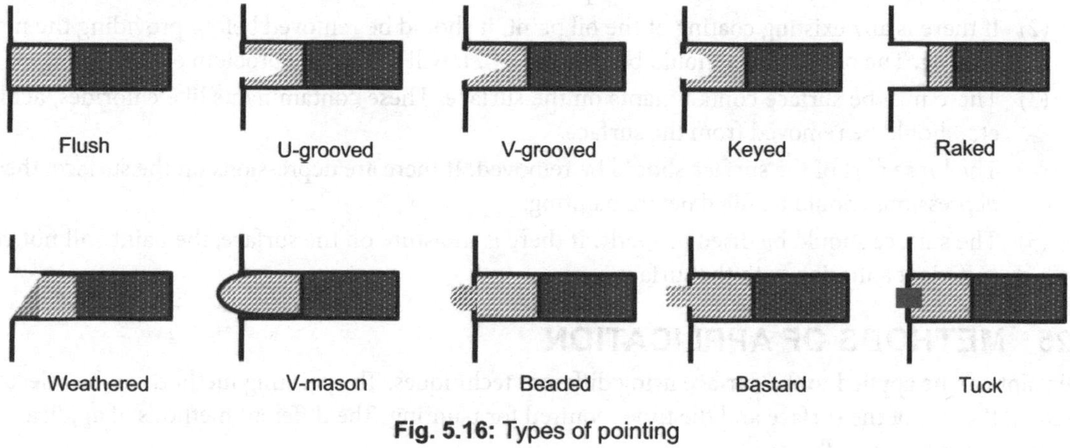

Fig. 5.16: Types of pointing

5.22.3 Procedure of Pointing

The pointing work is done with the masonry work of after the masonry work. If the pointing is carried out with masonry work, the pointing becomes strong. However, for better cleanliness, it can be done after the masonry work. The pointing work is done from the top to bottom. The surface is prepared before the start of the pointing work. The holes made during the scaffolding should be filled with mortar. The pointing work is carried out in the following sequence:

(1) The joints are scratched using a pointed hook. A 2 cm depth is scratched generally.
(2) The joints are cleaned properly. Water is sprayed on the joints.
(3) Mortar is filled and pressed in the joints. The mortar can be shaped in the required shape according to the type of pointing.
(4) Excess mortar is removed. It is scrapped from the joint and cleaned properly.
(5) After the setting of the pointing, the curing is done by sprinkling water. The curing should be done for 3 days.

5.23 PAINTING AND ITS NECESSITY

Painting may be defined as an oily coating that is provided on the surface of wood, masonry, iron, etc. A good paint should have sufficient covering capacity. It should be cheap and must dry quickly. The painting is required due to the following reasons:

(1) The painting protects the surface from the impact of weather. It prevents decay, corrosion, and rust.
(2) The paint protects the surface from the impact of sunlight and dust.
(3) It covers the defects of the structure.
(4) Good paint enhances the life of the structure.

5.24 SURFACE PREPARATION FOR PAINTING

The surface for the painting should be prepared carefully so that it can be ready to take the coating. The surface preparation is performed in the following steps:

(1) The condition of the surface should be assessed before beginning the painting work. The extent of rusting on the surface should be noted. The surface assessment helps in deciding the technique that can be used for the preparation of the surface.
(2) If there is any existing coating of the oil paint, it should be removed before providing the new coating. The old coatings should be stripped off. It will reduce the problem of corrosion.
(3) There may be surface contaminants on the surface. These contaminants like chlorides, acids, etc. should be removed from the surface.
(4) The loose part of the surface should be removed. If there are depressions on the surface, these depressions should be filled before painting.
(5) The surface should be dried properly. If there is moisture on the surface, the paint will not get sufficient adhesion with the surface.

5.25 METHODS OF APPLICATION

The paint can be applied to the surface using different techniques. The painting method can be selected based on the size of the surface and the time required for painting. The different methods of application of paint are explained as flows:

(1) **Brush:** Paintbrush is widely used for the application of paint. It is the most economical technique for the application of paint. The paintbrush applies the paint evenly on the surface. The fiber of the brush should be sufficiently elastic. The length of the paintbrush should be more than 1.5 times the width of the brush. It makes painting easy. The user should avoid using the new brush for the final coating.

(2) **Roller:** The painting by paintbrush takes time. So, the roller can be used. It consists of a cylinder that rolls while painting the surface. The roller for painting offers the following advantages:

➤ The paint rollers can cover a large area of painting in relatively lesser time.

➤ The paint rollers are user-friendly.

➤ Paint rollers can be used to give an artistic appearance to the surface.

➤ The rollers are easy to use. These can perform vertical painting easily.

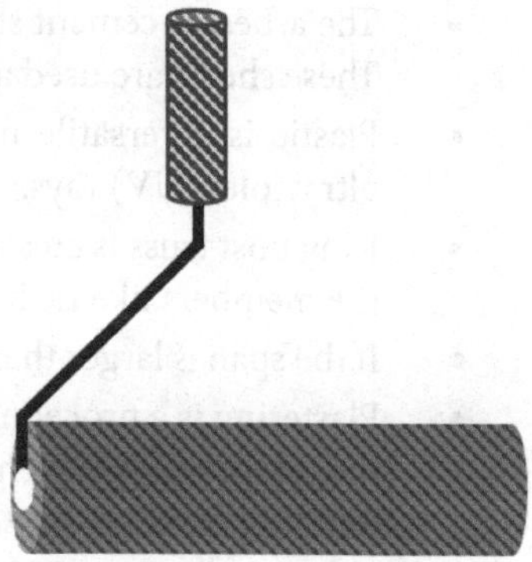

Fig. 5.17: Paint roller

The rollers are costlier as compared to the brushes. However, these prove economical as these can be replaced after degradation. Rollers can't be used in the corners. Figure 5.17 shows a paint roller.

(3) **Spray gun:** The painting of the surfaces can be done by the spray guns too. The spray guns spray the paint on the surface through the air. Air gun spraying uses large equipment for painting. The spray gun has the following features:

➤ The spray finish reduces the chances of stains and scratches as these are applied in thin layers.

➤ The spray gun technique is a time-saving technique for painting. It takes lesser time than brushing or roller.

➤ Spray guns can be used to apply paint in the corners and edges.

➤ The spraying technique is user-friendly and easy to apply.

➤ The quality of the painting by spraying is good. It provides consistency in the painting.

UNIT SUMMARY

- Floor is constructed into two parts i.e. base and flooring.
- The base provides the structural stability to the floor.
- The flooring or floor covering consists of the material that make the surface of the floor smooth.
- The floor covering can be of concrete, wood, tiles, paver blocks, etc.
- Ceramic tiles are prepared using clay while the vitrified tiles are prepared by mixing clay and silica.
- Chequered tiles have grooves and can fit each other.
- Skirting tiles are provided along the perimeter of the bottom of the wall. These tiles connect the floor and wall.

- Dado tiles are provided in the kitchen and washroom to protect the wall from the impact of water.
- The surface of the floor loses its shine with time. So, the floors need to be maintained periodically. The floors are polished and finished manually or mechanically.
- The permanent covering provided on the rooms to protect them from the impact of weather, sun, rain, storms, etc. is called a roof.
- The asbestos cement sheets are prepared by mixing 15% of the asbestos fiber into the cement. These sheets are used as the sheds of factories and workshops.
- Plastic is a versatile material that is used as a roof covering. It protects the building from ultraviolet (UV) rays.
- King post truss is created by the combination of triangular frames. A king post truss consists of the members like tie beams, rafters, struts, king posts, and ridge beams.
- If the span is larger than 9 m, the king post can't be used. In this case, a queen post truss is used.
- Plastering is a procedure in which a wet mixed material (cement/lime mortar) is used to cover the internal and external walls and rough patches on the surface of columns, beams, and other building parts.
- In Neeru finishing, lime paste or putty is used to render the cracks and level the uneven surface to obtain a smooth and leveled finish.
- Plaster of Paris is made by heating finely powered gypsum at 160°C - 170°C. It is white in color. It starts to get settle within 3 - 4 minutes of adding water.
- In the pebble finishing, the small pieces of stones having diameters from 10 mm to 20 mm are dashed on the surface of the wall.
- Wall cladding is the process of providing the layer of one material over the layer of other material to protect the building components from the impact of moisture.
- The process of coating the joints with mortar is known as pointing.
- Painting may be defined as an oily coating that is provided on the surface of wood, masonry, iron, etc. A good paint should have sufficient covering capacity.
- The paint can be applied by painting brush or roller.
- The paint can be sprayed on the surface too.

EXERCISES

Multiple Choice Questions

1. Which of the following is not a floor finish?

   (a) Marble (b) Kota
   (c) AC sheets (d) Granite

2. Vitrified tiles consist of:

   (a) Only clay (b) Only sand
   (c) Clay and silt (d) None of the above

3. Unglazed ceramic tiles are also called:

(a) Non-ceramic tile (b) Stone tiles
(c) Quarry tiles (d) None of the above

4. The tiles used in the bathroom should be:
(b) Porous (b) Impermeable
(c) Permeable (d) None of the above

5. In AC sheets the percentage of asbestos fibre is generally:
(a) 15% (b) 25%
(c) 30% (d) All of the above

6. The first coat of paint is called:
(a) First coat (b) Prime coat
(c) Primary coat (d) Final coat

7. Blistering is a defect in:
(a) Cement plaster (b) Paint
(c) Lime plaster (d) Wall cladding

8. The size of stones in the pebble finish is:
(a) 35 mm to 45 mm (b) 10 mm to 20 mm
(c) 25 mm to 35 mm (d) More than 45 mm

9. The process in which only joints are sealed with mortar is called:
(a) Painting (b) Pointing
(c) Semi-plastering (d) None of the above

10. Dry construction of the wall is done with the help of:
(a) Pebble finish (b) Plastering
(c) Plaster boards (d) All of the above

Answers of Multiple Choice Questions

1. (c), 2. (c), 3. (c), 4. (b), 5. (a), 6. (b), 7. (c), 8. (b), 9. (b), 10. (c)

Short and Long Answer Type Questions

1. Write a short note on Kota and marble flooring.
2. Write a brief note on chequered tiles and paver blocks.
3. Compare the concrete and wooden floor based on their performance.
4. Explain the process of laying the floor.
5. Explain the need for polishing of the floor.
6. Explain the difference between AC sheets, and GI sheets.
7. Explain the procedure of plastering.
8. Write a short note on Neeru finish.
9. Write a short note on stucco plaster.
10. Discuss the advantages of POP.
11. Explain the different defects in plastering.

12. Explain the different types of pointing.
13. Explain the process of preparation of surface for the painting.

KNOW MORE

Mangalore tiles are very famous and widely used in India. These tiles are known for their aesthetics and structural integrity but do you know Mangalore tiles have a great historical significance? The Mangalore tiles were discovered in 1860s. Georg Plebst, a German missionary, discovered the clay in Mangalore city that was used for the manufacturing of the Mangalore tiles. Scan the QR code to read this interesting article about the Mangalore tiles.

REFERENCES AND SUGGESTED READINGS

1. Chudley, R., Greeno, R. *Building Construction Handbook*. Elsevier, Butterworth-Heinemann, Amsterdam, 2007.
2. Punmia, B. C., et al. *Building Construction: An Elementary As Well As Advanced for Engineering Students*. Laxmi Publications, 2016.
3. Iano, J., Allen, E. *Fundamentals of Building Construction: Materials and Methods*. Wiley. 2019.
4. Gupta, D.V., Singh, V. *Building Construction Technology*. Asian Publishers, Muzzafarnagar, 2022.

APPENDICES

APPENDIX-A

NATIONAL BUILDING CODE OF INDIA

PART 4 FIRE AND LIFE SAFETY

1 SCOPE

1.1 This Code (Part 4) covers the requirements for fire prevention, life safety in relation to fire and fire protection of buildings. This Code (Part 4) specifies occupancy-wise classification, constructional aspects, egress requirements and protection features that are necessary to minimise danger to life and property from fire.

1.2 The provisions of this Part are applicable to,

a) all high rise buildings; and

b) special buildings, those are,

1) hotel, educational, institutional, business, mercantile, industrial, storage, hazardous and mixed occupancies, where any of these buildings have floor area more than 500 m² on any one or more floors;

2) educational buildings having height 9 m and above;

3) institutional buildings having height 9 m and above;

4) all assembly buildings;

5) buildings, having area more than 300 m² of incidental assembly occupancy on any floor; and

6) buildings with two basements or more, or with one basement of area more than 500 m².

unless otherwise mentioned specifically in the provisions.

NOTE — The owner of the building and parties to agreement, may however, decide to apply the provisions of this Part to buildings other than those given above.

2 TERMINOLOGY

For the purpose of this Part, the following definitions shall apply.

2.1 Assisted Evacuation — Strategy that exists during which a designated person or persons provide assistance, during an emergency, to another person(s) to leave a building or a specific part of the built environment and to reach a final place of safety.

2.2 Atrium — A large-volume space created by a floor opening or series of floor openings connecting two or more stories that is covered at the top of the series of openings and is used for purposes other than an enclosed stairway; lifts hoist-way; an escalator opening; or as a utility shaft used for plumbing, electrical, air conditioning, or communications facilities.

2.3 Authorities Concerned — An organization, office, or individual responsible for enforcing the requirements of a code or standard, or for approving or accepting equipment, materials, an installation, or a procedure.

2.4 Automatic Fire Detection and Alarm System — A system comprising components and sub-systems required for automatically detecting smoke, heat or fire initiating an alarm and other actions as appropriate. This system also includes manually operated electronic fire alarm (MOEFA) system.

NOTE — MOEFA system (with or without automatic fire detection and alarm system) includes all or some of the components such as manual call stations (initiating an alarm for fire and other actions as required), talk-back system and public address system.

2.5 Building — Any structure for whatsoever purpose and of whatsoever materials constructed and every part thereof whether used as human habitation or not and includes foundation, plinth, walls, floors, roofs, chimneys, plumbing and building services, fixed platforms, *Veranda*, balcony, cornice or projection, part of a building or anything affixed thereto or any wall enclosing or intended to enclose any land or space and signs and outdoor display structures. Tents, *Shamianahs*, tarpaulin shelters, etc, erected for temporary and ceremonial occasions with the permission of the Authority shall not be considered as building.

2.6 Building, Height of — The vertical distance measured in the case of flat roofs, from the average level of the ground around and contiguous to the building or as decided by the Authority to the terrace of the last livable floor of the building adjacent to the external wall; and in the case of pitched roofs, up to the point where the external surface of the outer wall intersects the finished surface of the sloping roof; and in the case of gables facing the road, the mid-point between the eaves level and the ridge. Where the building is located in a sloped terrain, height shall be determined from the lowest level (that is approachable by the fire service vehicles) to the terrace level. Architectural features serving no other function except that of decoration shall be excluded for the purpose of measuring heights.

APPENDIX-B

2.51 Public Way — A street, alley, or other similar parcel of land essentially open to the outside air, dedicated, or otherwise permanently appropriated to the public for public use and having a clear width and height of not less than 3 m.

2.52 Ramp — The construction, in the form of an inclined plane that is steeper than or equal to 1 : 20 (5 percent) from the horizontal, together with any intermediate landing, that makes it possible to pass from one level to another.

2.53 Refuge Area — An area within the building for a temporary use during egress. It generally serves as a staging area which is protected from the effect of fire and smoke.

2.54 Roof Exits — A means of escape on to the roof of a building, where the roof has access to it from the ground through alternative stair case or adjacent building.

2.55 Site (Plot) — A parcel (piece) of land enclosed by definite boundaries.

2.56 Smoke Barrier — A continuous membrane, or a membrane, where such membrane is designed and constructed to restrict the movement of smoke.

2.57 Smoke Compartment — A space within a building enclosed by smoke barriers on all sides.

2.58 Stack Pressure — Pressure difference caused by a temperature difference creating an air movement within a duct, chimney or enclosure.

2.59 Travel Distance — The distance to be travelled from any point in a building to a protected exit or external escape route or final exit measured along the line of travel.

2.60 Ventilation — Supply of outside air into, or the removal of inside air from an enclosed space.

2.61 Venting Fire — The process of facilitating heat and smoke to leave a building as quickly as possible by such paths so that lateral spread of fire and heat is checked, firefighting operations are facilitated and minimum fire damage is caused.

2.62 Visual Strobes/Flashing — It is an audio-visual fire alarm for alerting persons with hearing impairment with flashing light. The strobe frequency should be from 0.5 Hz to 4.0 Hz.

NOTE — Care should be taken to ensure that overlapping strobes do not combine to result in a higher frequency of flashing.

2.63 Volume to Plot Area Ratio (VPR) — The ratio of volume of building measured in cubic metre to the area of the plot measured in square metre and expressed in metre.

2.64 Water Based Systems

2.64.1 *Hydrant System* — A distribution system having a network of piping installed underground/above-ground around and/or through inside of a building with internal and/or external hydrants fitted with landing valves at regular intervals according to the occupancy. The distribution system is connected to water supply system for firefighting.

2.64.2 *Automatic Sprinkler System* — A system of water pipes fitted with sprinkler heads at suitable intervals and heights and designed to actuate automatically, control and extinguish a fire by the discharge of water.

2.64.3 *Automatic Water Spray Systems* — A special fixed pipe system connected to a reliable source of fire protection water supply and equipped with water spray nozzles for specific water discharge and distribution over the surface or area to be protected. The piping system is connected to the water supply through an automatically actuated deluge valve which initiates flow of water. Automatic actuation is achieved by operation of automatic detecting equipment installed along with water spray nozzles. There are two types of systems namely high velocity and medium velocity systems.

2.64.4 *Water Mist Systems* — A distribution system connected to a pumping and water supply system that is equipped with nozzles capable of delivering water mist to the part/entire enclosure or area, intended to control, suppress, or extinguish fire and is capable of meeting the specified performance requirements.

2.64.5 *Foam Protection System* — Firefighting systems where foam is made by mechanically mixing air with a solution consisting of fresh water to which a foaming agent (liquid concentrate) has been added. Firefighting foam is a stable aggregation of small bubbles of density lower than oil or water, and shows tenacious qualities for covering horizontal surfaces. There are three types of foam applications that is, low, medium and high expansion foams depending upon the application.

2.65 Wet Riser — An arrangement for firefighting within the building by means of vertical rising mains not less than 100 mm nominal diameter with landing valves on each floor/landing for firefighting purposes and permanently charged with water from a pressurized supply.

NOTE — For definition of other terms, reference shall be made to accepted standards [4(3)].

3 FIRE PREVENTION

3.1 Classification of Buildings Based on Occupancy

3.1.1 *General Classification*

All buildings, whether existing or hereafter erected shall be classified according to use or the character of occupancy in one of the following groups:

Group A	Residential
Group B	Educational
Group C	Institutional
Group D	Assembly
Group E	Business
Group F	Mercantile
Group G	Industrial
Group H	Storage
Group J	Hazardous

The details of each occupancy and example of buildings in each group are given in **3.1.2** to **3.1.10.**

3.1.1.1 *Minor occupancy*

This is purely incidental to operations in a main occupancy, which shall be considered as part of the main occupancy and shall be classified under the relevant group for the main occupancy.

3.1.1.2 *Mixed occupancy*

Where two or more types of occupancies intermingle in the same building, the entire building shall be treated as mixed occupancy and the same shall comply with **3.1.12**.

3.1.2 *Group A Residential Buildings*

These shall include any building in which sleeping accommodation is provided for normal residential purposes with or without cooking or dining or both facilities, except any building classified under Group C.

Buildings and structures under Group A shall be further subdivided as follows:

Subdivision A-1	Lodging and rooming houses
Subdivision A-2	One or two family private dwellings
Subdivision A-3	Dormitories
Subdivision A-4	Apartment houses
Subdivision A-5	Hotels
Subdivision A-6	Starred hotels

a) *Subdivision A-1 Lodging and rooming houses* — These shall include any building or group of buildings under the same management, in which separate sleeping accommodation on transient or permanent basis, with or without dining facilities but without cooking facilities for individuals is provided. This includes inns, clubs, motels and guest houses.

NOTE — A lodging or rooming house shall be classified as a dwelling in Subdivision A-2, if no room in any of its private dwelling units is rented to more than three persons.

b) *Subdivision A-2 One or two family private dwellings* — These shall include any private dwelling, which is occupied by members of one or two families and has a total sleeping accommodation for not more than 20 persons.

If rooms in a private dwelling are rented to outsiders, these shall be for accommodating not more than three persons per room.

If sleeping accommodation for more than 20 persons is provided in any one residential building, it shall be classified as a building in Subdivision A-1 or Subdivision A-4 as the case may be.

c) *Subdivision A-3 Dormitories* — These shall include any building in which group sleeping accommodation is provided, with or without dining facilities for persons who are not members of the same family, in one room or a series of closely associated rooms under joint occupancy and single management, for example, school and college dormitories, students, and other hostels and military barracks.

d) *Subdivision A-4 Apartment houses* — These shall include any building or structure in which living quarters are provided for three or more families, living independently of each other and with independent cooking facilities, for example, apartment houses, mansions and *Chawls*.

e) *Subdivision A-5 Hotels* — These shall include any building or group of buildings under single management, in which sleeping accommodation is provided, with or without dining facilities for hotels classified up to Four Star Category.

f) *Subdivision A-6 Starred hotels* — These shall include the hotels duly approved by the concerned authorities as Five Star and above hotels.

3.1.3 *Group B Educational Buildings*

These shall include any building used for school, college, other training institutions involving assembly for instruction, education or recreation for not less than 20 students.

Buildings and structures under Group B shall be further subdivided as follows:

Subdivision B-1	Schools up to senior secondary level
Subdivision B-2	All others/training institutions

a) *Subdivision B-1 Schools up to senior secondary level* — This subdivision shall include any building or a group of buildings under single management which is used for students not less than 20 in number.

b) *Subdivision B-2 All others/training institutions* — This subdivision shall include any building or a group of buildings under single management which is used for students not less than 100 in number.

In the case of temporary buildings/structures which are utilized for educational purposes, the provisions of **3.2.5.3** shall apply.

If residential accommodation is provided in the schools/institutions that portion of occupancy shall be classified as a building in Subdivision A-3.

3.1.4 *Group C Institutional Buildings*

These shall include any building or part thereof, which is used for purposes, such as medical or other treatment or care of persons suffering from physical or mental illness, disease or infirmity; care of infants, convalescents or aged persons and for penal or correctional detention in which the liberty of the inmates is restricted. Institutional buildings ordinarily provide sleeping accommodation for the occupants.

Buildings and structures under Group C shall be further subdivided as follows:

Subdivision C-1	Hospitals and sanatoria
Subdivision C-2	Custodial institutions
Subdivision C-3	Penal and mental institutions

a) *Subdivision C-1 Hospitals and sanatoria* — This subdivision shall include any building or a group of buildings under single management, which is used for housing persons suffering from physical limitations because of health or age and those incapable of self-preservation, for example, hospitals, infirmaries, sanatoria and nursing homes.

b) *Subdivision C-2 Custodial institutions* — This subdivision shall include any building or a group of buildings under single management, which is used for the custody and care of persons, such as children, convalescents and the aged who are incapable of self-preservation, for example, homes for the aged and infirm, convalescent homes and orphanages.

c) *Subdivision C-3 Penal and mental institutions* — This subdivision shall include any building or a group of buildings under single management, which is used for housing persons under restraint, or who are detained for penal or corrective purposes, in which the liberty of the inmates is restricted, for example, jails, prisons, mental hospitals, mental sanatoria and reformatories.

3.1.5 *Group D Assembly Buildings*

These shall include any building or part of a building, where not less than 50 persons congregate or gather for amusement, recreation, social, religious, patriotic, civil, travel and similar purposes, for example, theatres; motion picture houses; assembly halls; auditoria; exhibition halls; museums; skating rinks; gymnasiums; restaurants; places of worship; dance halls; club rooms; passenger stations and terminals of air, surface and marine public transportation services; and stadia.

Buildings under Group D shall be further subdivided as follows:

Subdivision D-1	Buildings having a theatrical or motion picture or any other stage and fixed seats for over 1 000 persons
Subdivision D-2	Buildings having a theatrical or motion picture or any other stage and fixed seats up to 1 000 persons
Subdivision D-3	Buildings without a permanent stage having accommodation for 300 or more persons but no permanent seating arrangement
Subdivision D-4	Buildings without a permanent stage having accommodation for less than 300 persons with no permanent seating arrangement
Subdivision D-5	All other structures including temporary structures designed for assembly of people not covered by Subdivisions D-1 to D-4, at ground level
Subdivision D-6	Buildings having mixed occupancies of assembly and mercantile (for example, shopping malls providing facilities such as shopping, cinema theatres, multiplexes and restaurants/food courts)
Subdivision D-7	Underground and elevated mass rapid transit system

a) *Subdivision D-1* — This subdivision shall include any building primarily meant for theatrical or operatic performances and which has a stage, proscenium curtain, fixed or portable scenery or scenery loft, lights, mechanical appliances or other theatrical accessories and equipment for example, theatres, motion picture houses, auditoria, concert halls, television and radio studios admitting an audience and which are provided with fixed seats for over 1 000 persons.

b) *Subdivision D-2* — This subdivision shall include any building primarily meant for use as described for Subdivision D-1, but with fixed seats up to 1 000 persons.

c) *Subdivision D-3* — This subdivision shall include any building, its lobbies, rooms and other spaces connected thereto, primarily intended for assembly of people, but which has no theatrical stage or permanent theatrical and/or cinematographic accessories and has accommodation for 300 persons or more, for example, dance halls, night clubs, halls for incidental picture shows, dramatic, theatrical or educational presentation, lectures or other similar purposes having no theatrical stage except a raised platform and used without permanent seating arrangement; art galleries, community halls, marriage halls, places of worship, museums, lecture halls, passenger terminals and heritage and archaeological monuments, pool and billiard parlours, bowling alleys, community halls, courtrooms, gymnasiums (without spectator seating), indoor swimming pools (without spectator seating), indoor tennis courts (without spectator seating).

d) *Subdivision D-4* — This subdivision shall include any building primarily intended for use as described in Subdivision D-3, but with accommodation for less than 300 persons with no permanent seating arrangements.

e) *Subdivision D-5* — This subdivision shall include any building or structure, permanent or temporary meant for assembly of people not covered by Subdivisions D-1 to D-4, for example, grandstands, stadia, amusement park structures, reviewing stands and circus tents, arenas, external swimming pools, tennis and similar type of courts.

f) *Subdivision D-6* — This subdivision shall include any building for assembly of people provided with multiple services/facilities like shopping, cinema theatres, multiplexes, restaurants/food court.

g) *Subdivision D-7* — This subdivision shall include any building or structure like example, underground or elevated railways.

3.1.6 *Group E Business Buildings*

These shall include any building or part thereof which is used for transaction of business for keeping of accounts and records and similar purposes, professional establishments, service facilities, etc. City halls, town halls, courthouses and libraries shall be classified in this group so far as the principal function of these is transaction of public business and keeping of books and records.

Buildings under Group E shall be further subdivided as follows:

Subdivision E-1	Offices, banks, professional establishments, like offices of architects, engineers, doctors, lawyers, post offices and police stations
Subdivision E-2	Laboratories, outpatient clinics, research establishments, libraries and test houses
Subdivision E-3	Electronic data processing centres, computer installations, information technology parks and call centres
Subdivision E-4	Telephone exchanges
Subdivision E-5	Broadcasting stations, T.V. stations and air traffic control towers

3.1.7 *Group F Mercantile Buildings*

These shall include any building or part thereof, which is used as shops, stores, market, for display and sale of merchandise, either wholesale or retail.

Mercantile buildings shall be further subdivided as follows:

Subdivision F-1	Shops, stores, departmental stores, markets (any with covered area up to 500 m^2)
Subdivision F-2	Shops, stores, departmental stores, markets (any with covered area more than 500 m^2)
Subdivision F-3	Underground shopping centres

Storage and service facilities incidental to the sale of merchandise and located in the same building shall also be included under this group.

3.1.8 *Group G Industrial Buildings*

These shall include any building or part of a building or structure, in which products or materials of all kinds and properties are fabricated, assembled, manufactured or processed, for example, assembly plants, industrial laboratories, dry cleaning plants, power plants, generating units, pumping stations, fumigation chambers, laundries, buildings or structures in gas plants, refineries, dairies and saw-mills, etc.

Buildings under Group G shall be further subdivided as follows:

Subdivision G-1	Buildings used for low hazard industries
Subdivision G-2	Buildings used for moderate hazard industries
Subdivision G-3	Buildings used for high hazard industries

The hazard of occupancy, for the purpose of the Code, shall be the relative danger of the start and spread of fire, the danger of smoke or gases generated, the danger of explosion or other occurrences potentially endangering the lives and safety of the occupants of the buildings.

Hazard of occupancy shall be determined by the Authority on the basis of the fire loads of the contents, and the processes or operations conducted in the building, provided, however, that where the combustibility of the material, the flame spread rating of the interior finish or other features of the building or structure are such as to involve a hazard greater than the occupancy hazard, the greater degree of hazard shall govern the classification.

For determination of fire loads and fire load density for arriving at the classification of occupancy hazard, guidance on calorific values of some common materials is given at Annex A.

A broad classification of industrial occupancies into low, moderate and high hazard classes is given at Annex B, for guidance. Any occupancy not covered in Annex B, shall be classified in the most appropriate class depending on the degree of hazard.

Where different degrees of hazard of occupancy exist in different parts of a building, the most hazardous of those shall govern the classification for the purpose of this Code, except in cases where hazardous areas are segregated or protected as specified in the Code.

a) *Subdivision G-1* — This subdivision shall include any building in which the contents are of such comparative low combustibility and the industrial processes or operations conducted therein are of such a nature that there is hardly any possibility for any self-propagating fire to occur and the only consequent danger to life and property may arise from panic, fumes or smoke, or fire from some external source.

b) *Subdivision G-2* — This subdivision shall include any building in which the contents or industrial processes or operations conducted therein are liable to give rise to a fire which will burn with moderate rapidity or result in other hazardous situation and may give off a considerable volume of smoke, but from which neither toxic fumes nor explosions are to be feared in the event of fire.

c) *Subdivision G-3* — This subdivision shall include any building in which the contents or industrial processes or operations conducted therein are liable to give rise to a fire which will burn with extreme rapidity or result in other hazardous situation or from which poisonous fumes or explosions are to be feared in the event of a fire.

3.1.9 *Group H Storage Buildings*

These shall include any building or part of a building used primarily for the storage or sheltering (including servicing, processing or repairs incidental to storage) of goods, ware or merchandise (except those that involve highly combustible or explosive products or materials), vehicles or animals, for example, warehouses, cold storages, freight depots, transit sheds, storehouses, truck and marine terminals, garages, hangars, grain elevators, barns and stables. Storage properties are characterized by the presence of relatively small number of persons in proportion to the area. Any new use which increases the number of occupants to a figure comparable with other classes of occupancy shall change the classification of the building to that of the new use, for example, hangars used for assembly purposes, warehouses used for office purposes, garage buildings used for manufacturing.

3.1.10 *Group J Hazardous Buildings*

These shall include any building or part thereof which is used for the storage, handling, manufacture or processing of highly combustible or explosive materials or products which are liable to burn with extreme rapidity and/or which may produce poisonous fumes or explosions for storage, handling, manufacturing or processing which involve highly corrosive, toxic or noxious alkalis, acids or other liquids or chemicals producing flame, fumes and explosive, poisonous, irritant or corrosive gases; and for the storage, handling or processing of any material producing explosive mixtures of dust which result in the division of matter into fine particles subject to spontaneous ignition. Examples of buildings in this class are those buildings which are used for,

a) storage, under pressure of more than 0.1 N/mm^2 and in quantities exceeding 70 m^3, of acetylene, hydrogen, illuminating and natural gases, ammonia, chlorine, phosgene, sulphur dioxide, carbon dioxide, methyloxide and all gases subject to explosion, fume or toxic hazard, cryogenic gases, etc;

b) storage and handling of hazardous and highly flammable liquids, liquefiable gases like LPG, rocket propellants, etc;

c) storage and handling of hazardous and highly flammable or explosive materials (other than liquids); and

d) manufacture of artificial flowers, synthetic leather, ammunition, explosives and fireworks.

NOTE — A list of hazardous substances giving quantities, for which or exceeding which owners handling such substances are required to be covered under *The Public Liability Insurance Act*, 1991, has been notified under the '*Rules on Emergency Planning, Preparedness and Response for Chemical Accidents*' by the Govt. of India, Ministry of Environment and Forests Notification No. G.S.R. 347(E) dated 01 August 1996.

3.1.11 *Mixed Occupancy*

In case of mixed occupancy, in so far as fire protection is concerned, all the occupancies/the entire building shall be governed by the most restrictive provisions of the Code among those applicable for individual occupancies. The provisions for life safety given in the Code for individual occupancy shall, however, apply to the respective occupancies. Exits in such mixed occupancy shall be arranged so as to ensure that means of egress is not decreased in the direction of egress travel.

Further, in such mixed occupancies, the occupancies are also required to be separated (horizontally and/or vertically as the case may be) by a 240 min fire resistance rating.

3.1.12 Where change in the occupancy of any building places it in a different group or in a different subdivision of the same group, such building shall be made to comply with the requirements of the Code for the new group or its subdivision.

3.1.13 Where the new occupancy of a building is less hazardous, based on life and fire risk, than its existing occupancy, it shall not be necessary to conform to the requirements of the Code for the new group or its subdivision.

3.1.14 A certificate of occupancy shall be necessary, as required under Part 2 'Administration' of the Code, before any change is effected in the character of occupancy of any building.

3.2 Fire Zones

3.2.1 *Demarcation*

The city or area under the jurisdiction of the Authority shall, for the purpose of the Code, be demarcated into distinct zones, based on fire hazard inherent in the buildings and structures according to occupancy (*see* **3.1**), which shall be called as 'Fire Zones'.

3.2.2 *Number and Designation of Fire Zones*

3.2.2.1 The number of fire zones in a city or area under the jurisdiction of the Authority depends upon the existing layout, types of building construction (*see* **3.3**), classification of existing buildings based on occupancy (*see* **3.1**) and expected future development of the city or area. In large cities or areas, three fire zones may be necessary, while in smaller ones, one or two may be adequate.

3.2.2.2 The fire zones shall be made use of in land use development plan and shall be designated as follows:

a) *Fire Zone No. 1* — This shall comprise areas having residential (Group A), educational (Group B), institutional (Group C), assembly (Group D), small business (Subdivision E-1) and mercantile (Group F) buildings, or areas which are under development for such occupancies.

b) *Fire Zone No. 2* — This shall comprise business (Subdivisions E-2 to E-5) and industrial buildings (Subdivisions G-1 and G-2), except high hazard industrial buildings (Subdivision G-3) or areas which are under development for such occupancies.

c) *Fire Zone No. 3* — This shall comprise areas having high hazard industrial buildings (Subdivision G-3), storage buildings (Group H) and buildings for hazardous uses (Group J) or areas which are under development for such occupancies.

3.2.3 *Change in the Fire Zone Boundaries*

When the boundaries of any fire zone are changed, or when it is intended to include other areas or types of occupancies in any fire zone, it shall be done by following the same procedure as for promulgating new rules or ordinances or both.

3.2.4 *Overlapping Fire Zone*

3.2.4.1 When any building is so situated that it extends to more than one fire zone, it shall be deemed to be in the fire zone in which the major portion of the building or structure is situated.

3.2.4.2 When any building is so situated that it extends equally to more than one fire zone, it shall be deemed

ANNEXURE A

Plan of House

ANNEXURE B

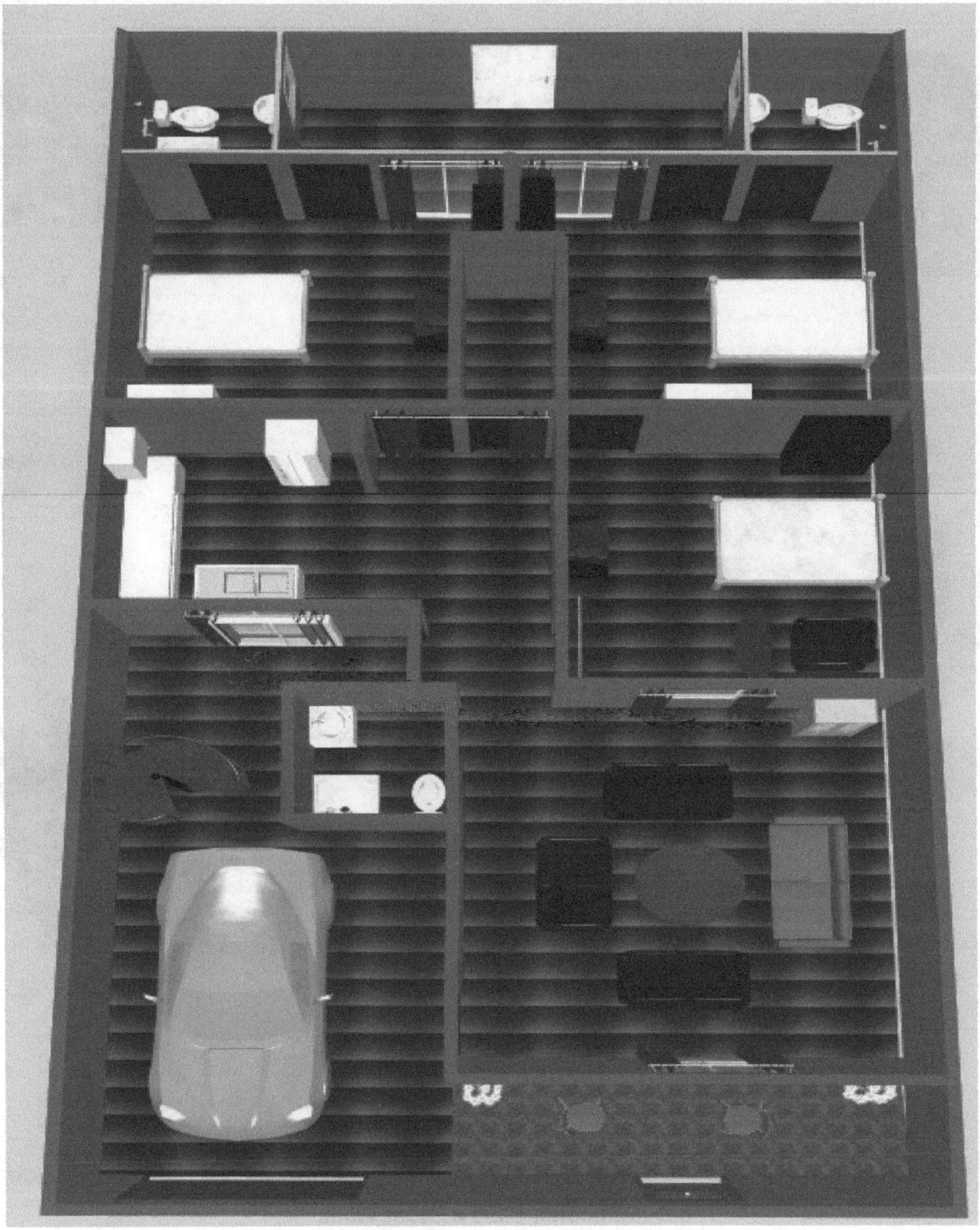

3 Dimensional View of House

ANNEXURE C

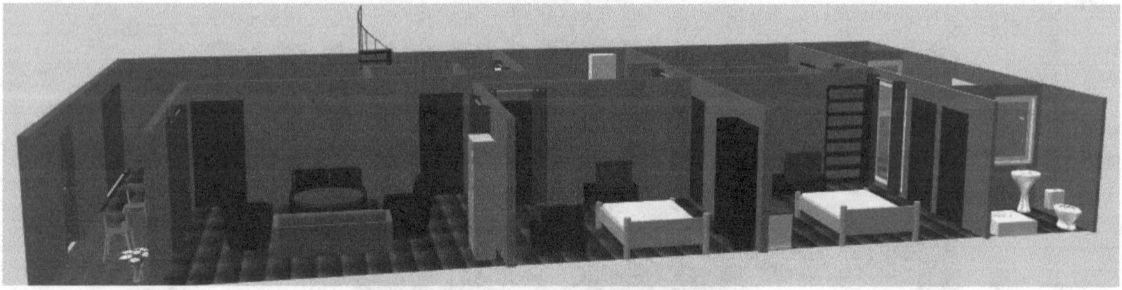

3 Dimensional Side View of House

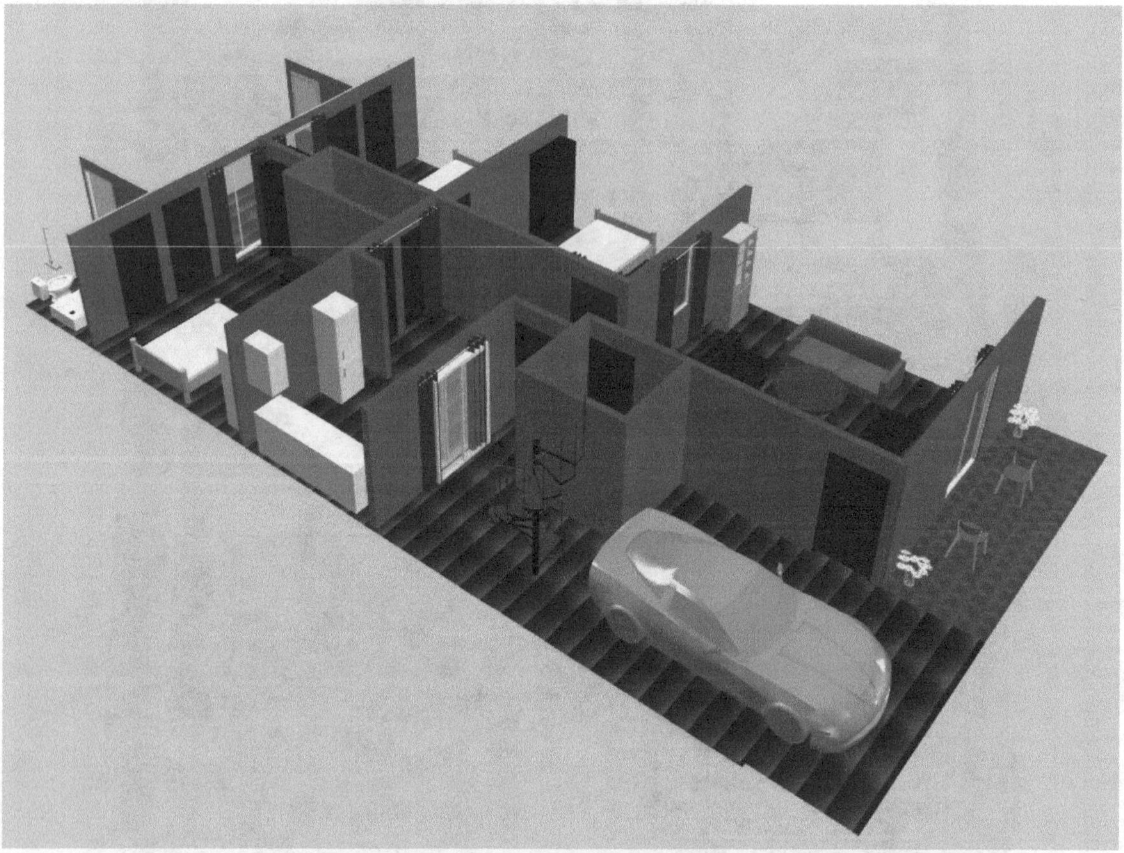

3 Dimensional View without Walls

REFERENCES FOR FURTHER LEARNING

1. Chudley, R., Greeno, R. *Building Construction Handbook.* Elsevier, Butterworth-Heinemann, Amsterdam, 2007.
2. Punmia, B. C., et al. *Building Construction: An Elementary As Well As Advanced for Engineering Students.* Laxmi Publications, 2016.
3. Iano, J., Allen, E. *Fundamentals of Building Construction: Materials and Methods.* Wiley. 2019.
4. Koskela, L., Huovila, P., & Leinone, J. (2002). Design Management in Building Construction: From Theory to Practice. *Journal of Construction Research, 03*(01), 1–16. https://doi.org/10.1142/S1609945102000035
5. Ebrahim, M. A.-B., Mosly, I., & Abed-Elhafez, I. Y. (2016). Building Construction Information System Using GIS. *Arabian Journal for Science and Engineering, 41*(10), 3827–3840. https://doi.org/10.1007/s13369-015-2006-1
6. Abuseif, M., & Gou, Z. (2018). A Review of Roofing Methods: Construction Features, Heat Reduction, Payback Period and Climatic Responsiveness. *Energies, 11*(11), 3196. https://doi.org/10.3390/en11113196
7. Fayad, F. al, Maref, W., & Awad, M. M. (2021). Review of White Roofing Materials and Emerging Economies with Focus on Energy Performance Cost-Benefit, Maintenance, and Consumer Indifference. *Sustainability, 13*(17), 9967. https://doi.org/10.3390/su13179967
8. Mallick, J., Ibnatiq, A. A., Kahla, N. ben, Alqadhi, S., Singh, V. P., Hoa, P. V., Hang, H. T., Hong, N. van, & Le, H. A. (2022). GIS-Based Decision Support System for Safe and Sustainable Building Construction Site in a Mountainous Region. *Sustainability, 14*(2), 888. https://doi.org/10.3390/su14020888
9. Yin, H., Li, Y., Zhang, D., Han, Y., Wang, J., Zhang, Y., & Li, A. (2022). Airflow pattern and performance of attached ventilation for two types of tiny spaces. *Building Simulation, 15*(8), 1491–1506. https://doi.org/10.1007/s12273-021-0876-6
10. dePolo, G., Walton, M., Keune, K., & Shull, K. R. (2021). After the paint has dried: a review of testing techniques for studying the mechanical properties of artists' paint. *Heritage Science, 9*(1), 68. https://doi.org/10.1186/s40494-021-00529-w

CO AND PO ATTAINMENT TABLE

Course outcomes (COs) for this course can be mapped with the program outcomes (POs) after the completion of the course and a correlation can be made for the attainment of POs to analyze the gap. After proper analysis of the gap in the attainment of POs necessary measures can be taken to overcome the gaps.

Table for CO and PO attainment

Course Outcomes	ATTAINMENT OF PROGRAMME OUTCOMES (1- Weak Correlation; 2- Medium correlation; 3- Strong Correlation)						
	PO-1	PO-2	PO-3	PO-4	PO-5	PO-6	PO-7
CO-1							
CO-2							
CO-3							
CO-4							
CO-5							

The data filled in the above table can be used for gap analysis.

INDEX

A

Assembly building 4
Ashlar masonry 62
Architrave 119
Apron 119
Aluminum window 122
AC sheets 150

B

Business building 4
Bracing 24
Box sheeting 25
Backhoe 3
Bulldozer 2
Foundation 29
Beveled closer 68
Bat 69
Bay window 124
Bifurcated stairs 136
Blistering 160

C

Cavity wall 9
Curved roof 13
Center line method 20
Course 60
Components of door 108
Combined footing 37
Coffer dams 50
Collapsible door 112
Cramped joints 66
Casement 119
Clerestory window 125
Corner window 125
Chajja 130
Chequered tile 145

D

Deep well 49
Dowelled joint 65
Doors 108
Door sizes 117
Dormer 126
Dog-legged staircase 134
Dado 147
Double coat plastering 156

E

Educational building 3
Earthwork 23
English bond 72
Excavation 23
Escalators 132
Eaves 152
Efflorescence 160

F

Framed structures 7
Flat roof 12
Face line method 21
Floating caissons 45
Flint walling 62
Flemish bond 74
Formwork 99
Flying shoring 97
Flush doors 115
Full paneled 120
Fixtures 128

G

Grillage footing 31
Glazed doors 112
Glazed window 121
Gable window 126
Granite 143
GI sheet 150

H

Hazardous building 5
Header bond 71
Hollow concrete block 89

I

Industrial building 5
Intersection 83

J

Job layout 18
Joints in stone 65
Joints in brick 69
Jambs 119

K

King closer 68
Kota finish 142
King post 151

L

Lap 69
Load bearing structures 5
Lintel 11
Louvered window 124
Lift 131

M

Mercantile building 4
Mezzanine floor 12
Mullion 109
Muntin 120

Metal stairs 138
Marble floor 143
Mangalore tiles 150

N

Nosing 132
Newel post 133
Neeru finish 156

O

Open well staircase 135

P

Partition wall 10
Parapet 14
Plinth materials 26
Pile foundation 42
Pier foundation 44
Pneumatic caissons 45
Polygon walling 62
Plywood formwork 101
Paneled doors 110
Pane 110
Partly paneled window 122
Paver blocks 145
Plastering 155
POP 157
Pebble finish 158
Plasterboard 158
Pointing 160
Paint roller 163

Q

Quoins 69
Quarter turn 135
Queen post 152, 154

R

Residential building 3
Rammer 28
Rake 28
Raft footing 41
Rubble masonry 59
Revolving doors 113
Rolling shutter 114
Reveal 119
Ramp 131
RCC stairs 138

S

Saddled joint 66
Sub-structure 7
Superstructure 9
Sill 10
Sloping roof 13
Site clearance 20
Spread footing 31
Stepped footing 32
Strip footing 34
Strap footing 40
Sheet pile 50
Stone masonry 57
Stretcher bond 71
Squint junction 84
Scaffolding 91
Suspended scaffold 93
Shoring 95
Sash 120
Sliding window 123
Spiral staircase 135
Shed 130
Stone stairs 137
Skirting 147
Stucco plaster 157
Sponge finish 158
Spray gun 163

T

Tee junction 78
Timbering and strutting 24
Tractor 20

V

Vertical sheeting 25
Vacuum well point 49
Vitrified tiles 144

W

Well footing 44
Windows 118
Window type 120
Wooden window 122
Winder 133
Wooden flooring 147
Wall cladding 159